Heat Pumps

Kazimierz Brodowicz
and
Tomasz Dyakowski

Translated by
Miroslaw L. Wyszyński

BUTTERWORTH
HEINEMANN

Butterworth-Heinemann Ltd
Linacre House, Jordan Hill, Oxford OX2 8DP

⟁ A member of the Reed Elsevier group

OXFORD   LONDON   BOSTON
MUNICH   NEW DELHI   SINGAPORE   SYDNEY
TOKYO   TORONTO   WELLINGTON

First published in Poland by Państwowe Wydawnictwo Naukowe 1990
© Państwowe Wydawnictwo Naukowe 1990

First published in Great Britain by Butterworth-Heinemann Ltd 1993

© Butterworth-Heinemann Ltd 1993

British Library Cataloguing in Publication Data
Brodowicz, Kazimierz
  Heat Pumps
  I. Title   II. Dyakowski, Tomasz
  III. Wyszyński, Miroslaw L.
  621.4

ISBN 0 7506 0611 8

Library of Congress Cataloguing in Publication Data
Brodowicz, Kazimierz.
  [Pompy ciepła. English]
  Heat pumps/Kazimierz Brodowicz and Tomasz Dyakowski; translated
  by Miroslaw L. Wyszyński.
  p.   cm.
  Includes bibliographical references and index.
  ISBN 0 7506 0611 8
  1. Heat pumps. I. Dyakowski, Tomasz. II. Title.
  TJ262.B7613   1993   93–10543
  621.402′5–dc20   CIP

Composition by Genesis Typesetting, Laser Quay, Rochester, Kent
Printed and bound in Great Britain by Redwood Books, Trowbridge

Contents

Preface

The natural and artificial environments both contain unlimited resources of energy at low exergy levels which cannot be utilized directly. Utilization of these resources is possible only by using heat pumps; hence the growing interest in heat pumps worldwide as well as in Poland.

The principles of operation of heat pumps are similar to those of refrigeration equipment. Heat pumps have, however, their own specific hierarchy of issues, so that in total they form a separate topic, which is the subject of this book. The book includes basic information on heat pumps, which is illustrated by the specific solutions as applied worldwide.

The book is addressed to engineers and students involved in municipal and industrial energy management, and also to anybody interested in the issues of energy economy.

The heat pump is virtually the only device enabling us to utilize the heat energy of sources which have low temperatures, and thus low exergy. There are significant reserves of such energy both in natural sources, e.g. atmospheric air or surface waters, and in artificial sources, such as various discharges of technological heat, which are particularly well suited to utilization. This energy is very cheap, or even free. Hence the interest in its utilization increases with the increasing costs or inconvenience of obtaining the highly exergetic energy. This highly exergetic energy has to be used anyway, and in not insignificant quantities, to drive the heat pump. A rationally working heat pump must increase the driving energy by adding the low exergetic energy to render the whole exercise profitable. This task is difficult to realize in practice. Because of that, although heat pumps have now been known for a century, only the last decade has seen the development of their theory and literature.

This book is addressed to graduate engineers and to students already having the basic background in engineering thermodynamics and theory of heat engines. Knowledge of basic refrigeration thermodynamics is particularly relevant, since the heat pump is theoretically similar to a refrigeration device, although its different objective puts different emphasis on individual problems. The authors aimed at presenting the comprehensive picture of issues relevant to heat pumps and at the same time they tried to discuss in more detail some important points related to their operation. Particular emphasis has been put on thermodynamics, with less space given to calculations or design of individual pieces of equipment. The authors were guided by the fact that the plant (apart from the compressor, the design of which is a separate issue) consists of heat and mass exchangers. Calculations and design of these are adequately developed and covered in the literature. The theoretical treatment is geared towards the needs of technological implementations, which are illustrated by many examples, constituting a substantial part of the work.

Our sincere thanks go to Professor Jan Berghmans (Katholieke Universiteit Leuven), Dr Jacek Kulesza, Professor Kazimierz Maczek and Professor Bogumil Staniszewski for help given in discussing the selection of material presented here.

Nomenclature

A surface area for heat or mass transfer (m^2): A_{ab}, of the absorber; A_{co}, of the condenser; A_{de}, of the desorber; A_{dp}, of the dephlegmator; A_{ev}, of the evaporator, A_{ew}, of the external surface area of the compressor's cylinder; A_{se}, of the solution heat exchanger; A_{ve}, of the vapour heat exchanger

B exergy (kJ): B_{iRhs}, the exergy transferred in the high-temperature heat source of the approximated cycle; B_D driving energy

B' rate of exergy flow (kW): B'_{de}, in the desorber

B_m magnetic field strength (T)

b specific exergy (kJ/kg): b_i, the specific thermal exergy (Equation 5.15); b_{iRhs}, in the high-temperature heat source of an ideal Rankine cycle

COP coefficient of performance (coefficient of effectivity of the heat pump related to the driving energy): COP_a, of an approximate cycle; COP_{aa}, of an absorption heat pump; $COP_{aa_{max}}$, the maximum value; COP_{ac}, of a compression heat pump; COP_{ar}, of a resorption heat pump; COP_{at}, of a heat transformer; COP_C, of the Carnot cycle; COP_i, of an ideal cycle (secondary subscripts as for COP_a); COP_{niR}, of the non-reversible Rankin cycle; COP_r, of a realized cycle (secondary subscripts as for COP_a); COP_t, of a theoretical cycyle (secondary subscripts as for COP_a)

c specific heat at constant pressure (kJ/(kgK)): c_{ex}, of the external heat source; c_{hs}, of the high-temperature heat source; c_L, of the liquid; c_{ls}, of the lean solution; c_{sr}, of the rich solution; c_{wL}, of the working fluid in liquid phase; c_{wV}, of the working fluid in gaseous phase

c cost function

D diameter (m): D_{cr}, of the rectification column

d diameter of Raschig rings (m)

đ symbol Phaffa: dq_{int}, internal heat; dw_{los}, loss work

E_D driving energy (kJ)

E'_{el} rate of the electrical energy flow (kW)

f ratio of the mass flow of the rich solution to the mass flow of working fluid, dimensionless

g free specific enthalpy (kJ/kmol), (see Appendix A.2.1)

H height of the rectification column (m)

HTU number of the heat transfer unit, dimensionless

I rate of flow of electric current (A)

i specific enthalpy (kJ/kg); i_L, for the liquid; i_v, for the vapour

k heat conductivity (W/(m^3 K))

l specific heat solution (kJ/kg); l_{ab}, in the absorber; l_{de}, in the desorber

M mass (kg)

m mass flow rate (kg/ms): m_{ext}, of the external heat source; m_{sl}, of the lean solution; m_{sr}, of the rich solution; m_{wV}, of the working vapour; m^*, mass flow rate per 1 kg/s of working fluid (Table 5.4)

N drive power (kW): N_{el}, electrical power; N_{Ht}, shaft power of the internal combustion engine; N_{ni}, of the non-isentropic compression; N_p, of the mechanical pump

n number of revolutions (rev/min^{-1})

NTU   number of (mass) transfer units, dimensionless

P pressure (MPa): P_{ab}, in the absorber; P_{co}, in the condenser; P_{cr}, critical pressure; P_{de}, in the desorber; P_{ev}, in the evaporator; P_{ext}, external pressure; P_{hs}, at the high-temperature heat source; P_{ls}, at low-temperature heat source; P_s, saturation pressure

PER   primary energy ratio (coefficient of effectivity of the heat pump cycle related to the primary energy)

Q heat, heat energy (kJ): Q_{ab}, in the absorber; Q_{ahs}, in the high-temperature heat source of an approximate cycle; Q_{als}, in the low-temperature heat source of an approximate cycle; Q_{Chs}, in the high-temperature heat source of the Carnot cycle; Q_{co}, in the condenser; Q_{de}, in the desorber; Q_{en}, environment heat; Q_{ev}, in the evaporator; Q_{ext}, external heat; Q_{hs}, in the high-temperature heat source; Q_{iRhs}, in the high-temperature heat source of an ideal Rankine cycle; Q_{ls}, in the low-temperature heat source; Q_{niRhs}, in the high-temperature heat source of a non-ideal Rankine cycle; Q_r, chemical reaction heat; Q_{re}, in the resorber.

Q' rate of heat flow (kW), subscripts as for the symbol Q, and additionally: Q'_j, the Joule's heat flow; Q'_k, the thermal conduction heat flow; Q'_P, the Peltier's effect heat flow; Q'_{se}, in the solution heat exchanger; Q'_{ve}, in the vapour heat exchanger

q heat exchanged in the individual exchangers of a heat pump installation, per one kilogram mass of working fluid, subscripts as for the symbol Q, and additionally: q_{ab}, in the absorber; q_{co}, in the condenser (Figure 2.4 appears in kWh/m^3); q_{de}, in the desorber; q_{dp}, in the dephlegmator; $q_{dp,min}$, for an infinite number of trays; q_{ev}, in the evaporator; q_{int}, heat produced from the dissipated work; q_{se}, of the solution heat exchanger; q_{ve}, of the vapour heat exchanger; q_{wc}, in the compressor cylinder

q' heat flux (kW/m^2)

R universal gas constant (kJ/(kmol K))

R electrical resistance (Ω)

r latent heat (kJ/kg): r_{co}, of condensation; r_{ev}, of evaporation (Equation 2.1, r appears in kJ/kmol)

S entropy (kJ/K): S_m, of magnetic field

S' entropy flow (kW/K)

s specific entropy (kJ/(kg K)): s_{cr}, of the critical point

T (absolute) temperature (K): T_{ab}, of the absorber; T_{bo}, the boiling temperature: $T_{bo,so}$, of the water solution; $T_{bo,w}$, boiling point for water at the same pressure as the solution; T_{co}, of the condenser; T_{cr}, critical temperature; T_{de}, of the desorber; T_{dp}, of the dephlegmator; T_{ev}, of the evaporator; T_{en}, of the environment; T_{hs}, of the high-temperature heat source; T_{ext}, of the external heat source; T_{ls}, of the low-temperature heat source; T_m, of the melting point (Table 2.7); T_m, mean temperature defined by Equation 4.16; T_s, saturation temperature; T_{se}, of the solution heat exchanger; T_{so}, of the heat source; T_{ve}, of the vapour heat exchanger; T_w, wall temperature; T_{wc}, of the compressor cylinder; $T_{w,ex}$, wall surface temperature outside; $T_{w,int}$, $T_{water,inlet}$, water inlet temperature; $T_{water,outlet}$, water outlet temperature

$\bar{T}$	(absolute) mean temperature (K), (Equations 1.10 and 4.9)
$T(x)$	(absolute) temperature distribution along heat transfer area: $T_{hs}(x)$, of the high-temperature heat source; $T_{ls}(x)$, of the low-temperature heat source
t	temperature (°C)
U	overall heat transfer coefficient (kW/m^{2}K): U_{ab}, of the absorber; U_{co}, of the condenser; U_{de}, of the desorber; U_{dp} of the dephlegmator; U_{ev}, of the evaporator; U_{se}, of the solution heat exchanger; U_{ve}, of the vapour heat exchanger
V	volume (m^3); V_{rel}, the relative displacement of a piston compressor, defined in Table 4.1
v	specific volume (m^3/kg): v_{sl}, of the lean solution; v_{sr}, of the rich solution; v_V, vapour; v_{wL}, of the working liquid
W	work (kJ): W_a, of the approximated cycle; W_C, of the Carnot cycle; W_{iR}, of the reversible Rankine cycle; W_{niR}, of the non-reversible Rankine cycle; W_r, of the practically realized cycle
w	work per one kilogram of working fluid (kJ/kg), subscripts as for the symbol W, and additionally: W_{loss}, loss work (Equation 4.4); W_{niR}, of the non-reversible Rankine cycle; W_{sl}, of the mechanical pump of the lean solution; W_{sr}, of the mechanical pump of the rich solution; W_{wL}, of the mechanical pump of the working fluid
w	relative mixture content of the soil, Figure 3.11
w	degree of dryness of vapour, dimensionless quantity
x	mass fraction of the working fluid in a two-component mixture, in the liquid; x_{sl}, of the lean solution; x_{sr}, of the rich solution
x	distance measured along the heat transfer area (m)
y	mass fraction of the working fluid in a two-component mixture, in the vapour; y_{wV}, of the working vapour
z	generalized mass fraction of the working fluid: z_{sl}, of the lean solution; z_{sr}, of the rich solution; z_{wV}, of the working vapour; z^*, mass fraction at the equilibrium line; z_{lim}, solubility limit
α	heat transfer coefficient (coefficient of heat transfer on one side of the heat transfer surface) (W/(m^{2}K))
α_{p-n}	Seebeck's coefficient for the p–n junction (V/K)
β	mass transfer coefficient (coefficient of mass transfer on one side of the transfer surface) (kg/(m^{2}s))
Δ	generally a difference; dimension depends on the quantities under consideration
ΔB	exergy difference (kJ)

Δb specific exergy difference (kJ/kg): Δb_{ahs}, in the high-temperature heat source of an approximate cycle

Δi enthalpy difference (kJ/kg): Δi_{cr}, of the chemical reaction (in Equations 1.1 and 1.2 Δi_{cr} appears in kJ/mol); Δi_d, heat of dissolution (in Table 2.7 Δi_d appears in kJ/mol); Δi_{is}, of the isentropic compression; Δi_r, of the realized cycle compression

ΔP pressure difference (MPa)

$\overline{\Delta P}$ mean driving force of elastic deformation (MPa) (Table 1.3)

ΔP_R negative deviation from Raoult's law (kPa)

Δs specific entropy changes (kJ/(kg K)): Δs_{HE}, Δ_{HP} of the dual Carnot cycle (Figure 5.1)

ΔS entropy difference (kJ/K)

$\Delta S'$ entropy flow (kW/K)

ΔT temperature difference associated with heat transfer (K): ΔT_{ab}, of the absorber; ΔT_{co}, of the condenser; ΔT_{de}, of the desorber; ΔT_{dp}, of the dephlegmator; ΔT_{ev}, of the evaporator; ΔT_{hs}, of the high-temperature heat source; ΔT_{ls}, of the low-temperature heat source; ΔT_{se}, of the solution heat exchanger; ΔT_{ve}, of the vapour heat exchanger

ΔT_{ln} logarithmic mean temperature difference (K)

$\Delta T(x)$ local temperature difference (K): $\Delta T_{hs}(x)$, of the high-temperature heat source; $\Delta T_{ls}(x)$, of the low-temperature source

$\overline{\Delta T}$ mean temperature difference (K), defined by Equation 1.10

Δv specific volume difference (m³/kg), (Table 2.4)

ΔV volume difference, (m³)

ΔU voltage difference, (V)

Δz_{ln} logarithmic mean concentration difference, dimensionless

δ generally an increase, dimension depends on the quantities under consideration

δB loss of exergy (kW)

$\delta B'$ loss of exergy flow (kW): $\delta B'_{cv}$, in the control valve; $\delta B'_{di}$, in the diffuser; $\delta B'_{en}$, in the expansion nozzle; $\delta B'_{mc}$, in the mixing chamber

$\delta b'$ loss of specific exergy flow (kW/kg)

δb loss of specific exergy (kJ/kg): δb_{co}, in the condenser

δP pressure drop (MPa): δP_{hs}, in the high-temperature heat source; δP_{iv}, in the compressor inlet valve; δP_{ls}, in the low-temperature heat source; δP_{ov}, in the compressor outlet valve

δS increase of entropy (kJ/K): $\delta S_{\Delta P}$, associated with the pressure drop in flow (in Table 1.3 associated with volumetric deformation); $\delta S_{\Delta T}$, associated with the temperature drop in heat transfer, $\delta S_{\Delta U}$, associated with electrical energy

$\delta S'$ increase of entropy flow (kW/K)

δs increase of specific entropy (kJ/(kg K)): δs_{co}, in the condenser; δs_{ev}, in the evaporator; δs_{ex}, in the expansion valve; δs_{iv}, at the inlet to the compressor; δs_{ov}, at the outlet to the compressor; δs_{wc}, in the compressor cylinder

δT temperature changes of substance in the source: δT_{ab}, in the absorber; δT_{co}, in the condenser; δT_{de}, in the desorber; δT, in the dephlegmator;

	δT_{ev}, of the evaporator; δT_{hs}, in the high-temperature heat source; δT_{ls}, in the low-temperature heat source; δT_{se}, in the solution heat exchanger; δT_{ve}, in the vapour heat exchanger
δw_{loss}	loss work (kJ)
ε	efficiency of a heat exchanger: ε_{se}, of the solution exchanger; ε_{ve}, of the vapour exchanger, dimensionless quantity
ε^*	porosity of soil: ratio of the volume of voids in soil to its total volume (Figure 3.11), dimensionless quantity
η	efficiency of converting one form of energy into another: η_{HE}, heat energy into mechanical energy (resulting from the Second Law of Thermodynamics); η_{el}, electrical energy into mechanical energy; η_{is}, of the isentropic compression; η_m, mechanical energy into mechanical energy; $\eta_{m,di}$, of the diffuser; $\eta_{m,en}$, of the expansion nozzle; $\eta_{m,sl}$, of the mechanical pump of the lean solution; $\eta_{m,sr}$, of the mechanical pump of the rich solution; $\eta_{m,wL}$, of the mechanical pump of the working fluid; η_{ic}, isentropic compression
η	efficiency of mass transfer: η_{ab}, in the absorber, η_{de}, in the desorber
η^*	exergetic efficiency: η_a^*, of an approximate cycle; η_{aa}^*, of an absorption heat pump; η_{ac}^*, of a vapour compression heat pump; η_{at}^*, of a heat transformer; η_r^*, of a realized cycle (secondary subscripts as for η_a^*); η_t^*, of a theoretical cycle (secondary subscripts as for η_a^*); η_{HE}^*, general efficiency of a heat pump; η_{HP}^*, of a heat engine, η_{niR}^*, of a non-ideal reverse Rankine cycle
π_{p-n}	Thomson's constant (V)
ρ	specific density (kg/m^3): ρ_L, of the liquid; ρ_V, of the vapour
ρ_{el}	specific resistivity (Ω/m^2)
μ	dynamic viscisity coefficient (kg/(ms))
σ	degree of regeneration, dimensionless quantity (Equation 4.28)
σ	surface tension (N/m)
$\varphi_{n,m}$	coefficient of cycle comparison for a compression heat pump, first subscript denotes a cycle n, having larger number of losses (irreversible processes), second a cycle m which takes into account fewer losses or even an ideal cycle: $\varphi_{a,C}$, $\varphi_{a,iR}$, $\varphi_{a,niR,iR}$, $\varphi_{niR,C}$, $\varphi_{niR,iRa,r}$, $\varphi_{r,C}$, $\varphi_{r,iR}$, $\varphi_{r,niR}$
ψ	decisive variable (Table 4.8)

Chapter 1

Introduction

1.1 Historical notes

The first theoretical information on the possibility of using a heat pump for heating was given in 1852 by W. Thomson (Lord Kelvin). He described an open cycle employing air, with a compressor and two water tanks working as the high- and low-temperature heat sources. Despite many attempts at practical exploitation of this idea, it remained unrealized for many years. The 1880s saw the beginnings of development of refrigeration technology, utilizing the vapour compression cycle working on ammonia. Using the elements of such an plant, T.G. Haldane[1] in 1928 built an installation for heating his own home. He was a great enthusiast of the heat pump, but did not manage to popularize it any further.

The first practically implemented vapour compression cycle heat pumps, fully correct technically and run continuously over many years, were of relatively high-power. A heat pump having power of 1050 kW and coefficient of performance $COP_{rc} = 2.5$ was installed in 1930 at the Southern California Edison Co. in Los Angeles. The Town Hall in Zurich was equipped in 1938 with a heat pump of 175 kW power and COP_{rc} of 2, and in 1942 the Technical University ETH in Zurich installed a heat pump of 7 MW power and $COP_{rc} = 3$.

In the 1940s, with the advent of the widespread use of organic working fluids, low-power heat pumps for heating individual houses appeared on the market. The first installations were implemented in the United States, because of advantageous climatic conditions there. In winter the heat pump was used for heating, and in the summer as an air conditioner. By the end of the 1970s, approximately 850 000 heat pumps were installed. At the time of highest demand for small heat pumps, which happened in the middle 1980s, 30% of all new individual houses in the United States were equipped with heat pumps. Heat pumps are also used in countries having quite different climatic conditions, such as Norway, Sweden, Germany, France and Italy. Since 1990, heat pumps have also been installed in Poland, although still in quite limited numbers.

The interest in the application of compression heat pumps in municipal economies increased very significantly in the 1980s. It has been estimated that by the end of 1986 approximately 40 large heat pumps had been installed, supplying in total 600 MW of heat energy (power).

In a vapour compression cycle, heat pump electrical energy is converted into heat energy. The electrical form of energy is easy to transmit, hence the appeal of small heat pumps, because of their wide installation possibilities. Alas, the conversion of one form of energy into another is associated with irreversible losses of exergy. Thus the incorporation of a heat pump early in the energy transformation chain can yield significant gains, and that is why heat energy is used directly to drive the pump.

Several technical solutions are possible, but the sorption heat pumps provide a quite attractive and practically proven alternative. They do not have any moving parts (except for the liquid pumps), which accounts for their great reliability. They are implemented in several versions, and in particularly advantageous conditions their capital cost is recovered in less than a year.

The use of sorption cycle heat pumps is becoming ever more widespread, both in industry and in the municipal economies. Heat power in the high-temperature heat source can achieve tens of megawatts. They can also be used when, for environmental protection reasons, the driving heat is supplied by the hot discharge water, which is thus cooled. In such applications the sorption cycle heat pump used as a heat transformer has no competition, and on top of the energy effects it has an important ecological function.

1.2 Applications of heat pumps

As already mentioned, the majority of heat pumps in mass use are employed in heating individual houses. The power of such heat pumps ranges from several to a dozen or so (only rarely up to 50) kilowatts at the level of the high-temperature heat source. They are largely vapour compression cycle types, usually fitted with piston compressors driven by electric motors. The majority draw heat from atmospheric air. Often these pumps are combined with a different, electrical or central heating system, used in series or in parallel. A large proportion of these pumps work as heat pumps during the winter, and as air-conditioning appliances during the summer.

Heat pumps of small power, in the range of several kilowatts, can be used to provide small quantities of domestic hot water. In such an application a heat pump provides adequate temperatures and high-efficiency, particularly in the summer.

Heat pumps with power in the range of a dozen or so kilowatts are used for heating larger interiors, such as trade storage rooms, or for heating water in swimming pools. These are usually vapour compression cycle types, fitted with piston compressors, with the low-temperature heat source provided by atmospheric air or alternatively by surface or underground waters.

Larger heat pumps (vapour compression cycle type), with powers of several megawatts, require the use of flow-compressors, which are difficult to manufacture and are not made in Poland. For this reason the sorption cycle type heat pumps can be very attractive in such applications. These installations are generally used for heating housing estates or supplying them with hot water. Increasingly they are also used in industry.

Particularly advantageous conditions for installing heat pumps exist in several cases:

1. Where there already exists a source of discharge heat at a temperature too low (but still higher than the environmental temperature) to be utilized without a

heat pump, and where at the same time and place there is a requirement for heat energy;

2. Where there is a requirement for both heating (or, to be more precise, for energy released by the heat pump at the high-temperature heat source level) and for cooling effects (i.e. for the energy absorbed at the low-temperature heat source level for cooling purposes) – these requirements can be either simultaneous or seasonally changing;

3. Where in an industrial plant there is a large energy flow which can be reversed by using a heat pump (so that energy will no longer be supplied at the inlet and discharged at the outlet). Examples of such cases are in air conditioning, dryers, evaporators, rectifying plant, installations for washing equipment or bottles the in food industry;

4. Where in an industrial installation there already exists an extensive system of heat regeneration and the use of a heat pump can improve this system;

5. Where the energy is being transmitted over significant distances so that by using a heat pump (particularly a heat transformer) at the receiving end, one can significantly reduce the capital investment costs.

1.3 Arguments for and against the use of heat pumps

As the heat pump is a well-known device and is not a novelty even in Poland (because of its similarity to refrigeration equipment), one can put together the points for and against its application. These points can be of various kinds – thermodynamic, technological, energetic, ecological and economic – and can take the form of an analysis which should help to judge whether the use of a heat pump is appropriate. An economic analysis can only be performed for a definite set of conditions, and the steps of analysis can be illustrated using an example of a domestic compression-type heat pump. Let us take a family house which is characterized by a certain energy requirement and which can be heated in several ways: by burning solid fuel (coal) or gas, by electricity or by using a heat pump. On putting together the costs of fuel and an installation for its burning, the costs of electrical energy and the electric heating installation, the costs of installation and running of the heat pump, one can assess the conditions in which the use of a heat pump will yield measurable savings. For example, it would be installed if the capital investment costs could be returned within two years.

The arguments supporting the case for using heat pumps and the necessity of their further development can be listed as follows:

1. We are all facing the necessity of saving energy or minimizing its use due to global limitations; without these savings it is difficult to imagine any further expansion of the economy (on both the global and national scale). The heat pump can play an important part in this context.

2. It is generally known that a heat pump can be used for heating purposes and that it can compete with other devices; less well known is the fact that a heat pump can become an important element of an energy system, improving its efficiency, reducing the investment costs of newly installed plant, or becoming an

important element in the modernization of existing installations. In this context a heat pump might be without competition.

3. A heat pump might work as a heating device (which can easily be replaced by a different system) and as a cooling device (which cannot be replaced by another system), so that when there is a simultaneous requirement for heating and cooling a heat pump is excellent.
4. A vapour compression cycle heat pump driven by an electric motor does not pollute the (local) environment, so that it can be used in locations requiring minimum pollution (e.g. spas), as well as in locations which are already heavily polluted (e.g. industrial centres), where one cannot install any more heating systems of other kinds.

It is quite impossible to present absolute arguments against using heat pumps, particularly as they are being developed in the technologically leading countries. It also does not seem probable that in the longer term the price of energy will show a tendency to drop, and hence the necessity to save energy will certainly become more urgent. For this reason the following arguments 'against' using the heat pumps have limited application:

1. In Poland one cannot count on the mass application of vapour compression cycle heat pumps for heating individual family houses, as is the case, for example, in the USA (where heat pumps are often combined with air-conditioning devices and are used as coolers in the summer), because of the different climate and different urban development.
2. A heat pump used for heating apartments compares quite poorly (in energy terms) with the district heating system quite commonly used in Poland in large urban centres.
3. A heat pump is a relatively poor competitor in economic terms with coal-fired heating systems.
4. Vapour compression cycle heat pumps, with power in the order of several megawatts, are better suited to the largely centralized energy management systems. The energy management considerations are technically and economically important factors.
5. Each installation, particularly of a large heat pump (based on the sorption cycle), has to be designed and built individually; such plant is more complex and less traditional than other heating installations, hence there is reluctance to run it, and also fears of facing possible shortages of qualified staff, leading to a lack of interest from the manufacturing and heating industries.
6. Lack of working fluids suitable for practical implementation of heat pump cycles, which would guarantee suitably good parameters.

The information relating to desirable conditions for the use of heat pumps and to their advantages and disadvantages, collected above, takes into account various aspects: thermodynamic, ecological, economic. Surely the last two are the most important, but in Poland these are difficult to define unambiguously. It seems, therefore, that the following reasoning should govern the analysis of the usefulness of heat pumps. On one hand, there are quite unambiguous thermodynamic conditions relating to the usefulness of heat pump installation; these conditions are

Table 1.1 *Suitabilities of heat pumps for various applications*

No.	Task and power of the heat pump	Temperature of sources	Degree of suitability of various types of heat pumps*			
			Poor	Average	Good	Very good
1	Heating of individual dwellings by heat taken from atmosphere Q_{hs} 10 kW	$T_{hs} = 330$ K $T_{ls} = 270$ K	C 1.5	A_e 3.0	A_s 3.5	
2	Reversing the heat flux 0.25–1 MW					
	Evaporators	$T_{ls} = 400$ K			C 5	G 9
	Rectification towers	$T_{ls} = 340$ K				F 0.5
	Equipment washing plant				A_e 3.0	F 0.5
	Dryers	$T_{hs} = 320$ K $T_{ls} = 300$ K		D 1.4	A_e 3.5	H 3.5
3	Recovery of waste heat sources for technological, municipal and other use 1–5 MW					
	Production of technological steam	$T_{hs} = 400$ K $T_{ls} = 340$ K			D 1.4	F 0.5
	Co-operating with central heating or combined heat and power	$T_{hs} = 350$ K $T_{ls} = 320$ K			B 3.5	D 1.4
	Recovery of heat from condensation, municipal water	$T_{hs} = 320$ K $T_{ls} = 300$ K			C 1.5	E 4.5

*Type of heat pump: A, compressor based, with piston compressor; A_e, driven by electrical energy; A_s, driven by internal combustion engine or a gas engine; B, compressor based, with a through-flow type of compressor; C, compressor based, with ejector compression; D, sorption based, absorption type; E, sorption based, resorption type; F, heat transformer; G, utilizing the compression of vapours; H, with a through-flow compressor and air cycle. Figures given underneath the letter symbols for various heat pumps are approximate values of the COPs.

of fairly universal character. On the other hand, one has access to the published data relating to applications of heat pumps in heating management in Germany, Japan and England. The conclusion from this information [2] is that heat pumps are so cost-effective that the capital costs are sometimes recovered within less than a year. These data also yield information on the various types of heat pumps used in different conditions. Starting with this information, we have prepared Table 1.1, which gives a compilation of the suitabilities, on an arbitrary scale, of various types of heat pumps.

1.4 Thermodynamic principles of heat pump operation

1.4.1 Ways of accomplishing the heat pump cycle

The basic task of a heat pump, conveyance of heat from the low-temperature heat source to the high-temperature heat source (Figure 1.1), can be accomplished in various ways. The most important one is the vapour cycle (anticlockwise), but a gas cycle is also possible. One can also utilize the heats of dissociation and synthesis in

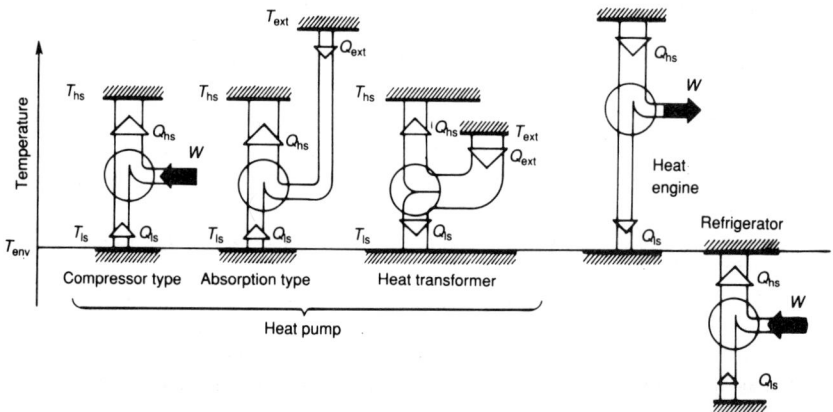

Figure 1.1   *The principle of operation of some more important heat pumps and their comparison with a heat engine and a refrigerator*

chemical reactions, as well as heats of various effects: Ranque's, thermoelectric or magnetic. One should, however, stress the special importance of vapour cycle heat pumps; the use of other types is only justified by theoretical possibilities. Only the vapour cycle heat pumps are currently in use, and so far only these can compete with the traditional heating systems in economic terms. One should add that the vapour cycle heat pumps are implemented in different variants of design. Therefore some further chapters (Chapters 4, 5 and 6) refer to these pumps, and the information necessary to understand the principles of their operation is given there. On the other hand, heat pumps utilizing the gas cycle, thermoelectric effect, magnetic effect or Ranque effect, as well as those utilizing the heat effects of chemical reactions, are not described further. For this reason a bit more space is given to these pumps here, including their quantitative characteristics.

The vapour cycle of a heat pump consists of two isobaric changes of phase effected at different pressures, compression of vapour between these pressures and expansion of the liquid. Taking into account the possible variants of effecting these processes (in different devices), one can propose several ways of accomplishing the vapour cycle, which are illustrated in Figure 1.2, while Figure 1.3 shows some more important heat pumps following these implementation versions.

Vapour compression, strictly speaking the kind of energy used to do the compression work, is a particularly important issue in the operation and classification of heat pumps. This energy can be either mechanical work or heat. Mechanical work is delivered directly to the compressor shaft and in such a case we are not interested in the process of obtaining this work (e.g. in an electric power station or in an internal combustion engine). If, however, the heat energy is supplied, it has to be converted into work first, and only then can it be passed on to do the compression work. Because of this, a heat pump cycle realized by supplying the heat is in effect a double cycle, including, apart from an actual (anticlockwise) heat pump cycle, a heat engine cycle (clockwise). This heat engine cycle can be implemented by using various working fluids. When a single-component working fluid is used, the heat energy supplied in the evaporator is converted in the injector to kinetic energy, which is next transformed in the diffuser into the vapour compression work in the actual heat pump cycle. When a combination of a working fluid and sorbent is used, the heat engine cycle in a sorption cycle heat pump is realized between the absorber and the desorber. The delivered heat (the difference between Q_{ext} and Q_{hs2}) is converted into the increase of enthalpy of the fluid flowing between the absorber and the desorber. In effect, the pumping of fluid in its liquid phase is accompanied by the compression of its vapour in the actual heat pump cycle, so that work is delivered to this cycle.

The condensation, defined generally as a process of increasing the density of matter, can be accomplished by discharge of heat at constant pressure during the processes of condensation or absorption. A change opposite to condensation, defined as decreasing the density of matter, can be accomplished by evaporation or desorption.

The pressure drop in a region where the fluid is in two physical phases, but close to the liquid state, can be accomplished by expansion, accompanied by a discharge of work to the outside, or by throttling in the expansion valve.

The simplest operational principle to explain is that of a heat pump to which mechanical work is delivered. The phase changes occur in the condenser and evaporator, and expansion of the liquid occurs in a valve. The thermodynamic cycle of such a pump, and the installation which makes it possible, are shown in Figure 1.3a using an anticlockwise Carnot cycle. The Carnot cycle is an ideal cycle independent of the working fluid; therefore it is not the best comparative cycle, but it enables consistent illustration of the vapour cycle in all heat pumps. Later on in the book we will introduce better comparative cycles.

More difficult to describe are heat pumps, which in their installation include a heat engine. The heat engine cycle can be integrated to varying degrees with the

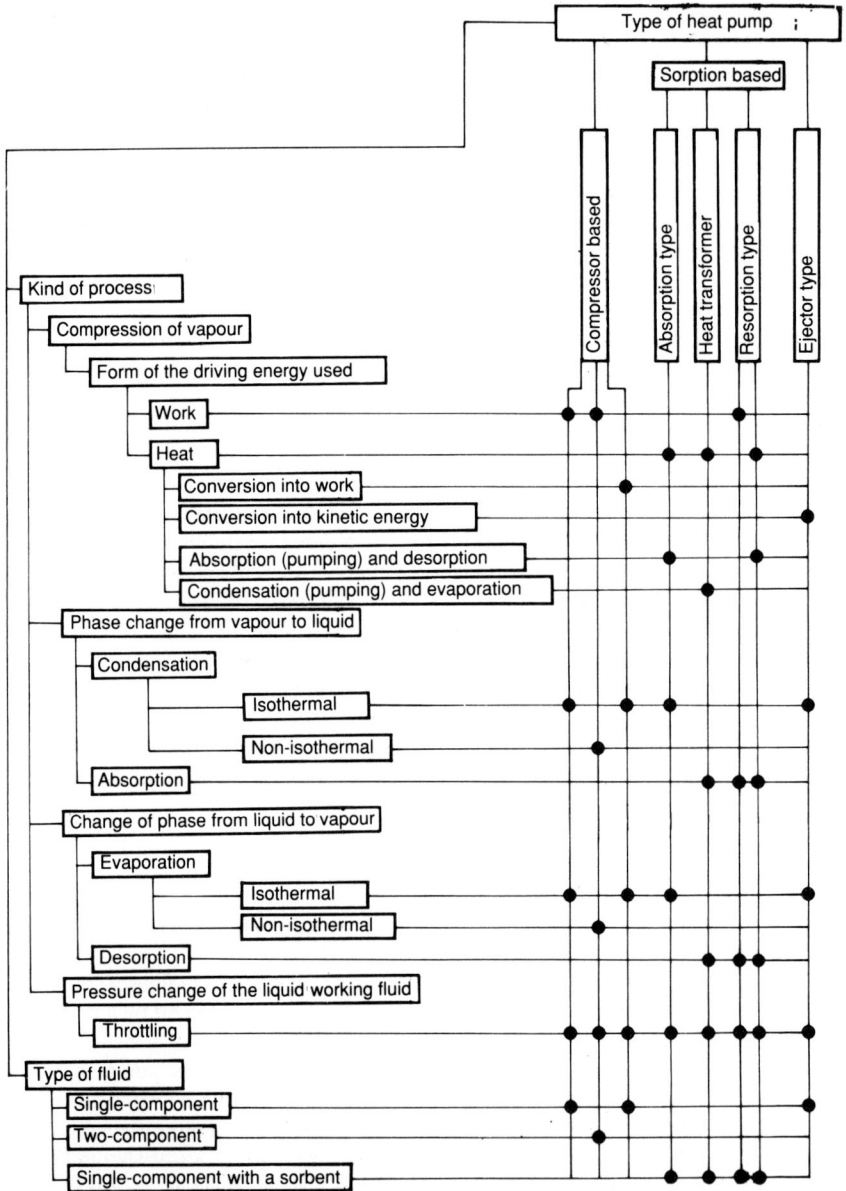

Figure 1.2   *Various types of heat pumps, with different implementations of individual processes in the cycle*

actual heat pump cycle. When this integration is only superficial, it is easy to separate the actual heat pump cycle from the heat engine cycle and to illustrate both of these by the comparative Carnot cycles (Figure 1.3b,c). When the integration is deeper, particularly in a sorption cycle heat pump, separation of the two cycles is more complicated. This complication results from the fact that for a two-component and two-phase system the thermodynamic liquid–vapour equilibrium is not a line any more, but, according to the Gibbs rule, it has to be represented by a surface. Thus the cycle has to be represented not on a plane, but in three-dimensional space. An attempt at such interpretation was made for an absorption cycle heat pump and for a heat transformer, shown in Figure 1.3d and 1.3e, by illustrating their cycles in the three-dimensional space $T–s–z$ (see Figures 1.6b and 1.7b).

Summing up, we have introduced seven types of heat pump utilizing the vapour cycle, and if we add the two possible variants of gas cycle heat pumps and the heat pumps utilizing the heat of chemical reaction and the Ranque, thermoelectric, electrodiffusion and magnetic effects, we can specify 14 types of heat pump:

1. Vapour compression cycle heat pump with a compressor and a single-component working fluid
2. Vapour compression cycle heat pump with a compressor and a two-component working fluid
3. Absorption cycle heat pump
4. Absorption cycle heat transformer
5. Absorption cycle heat pump driven by mechanical energy
6. Absorption cycle heat pump driven by heat energy
7. Heat pump utilizing vapour compression
8. Gas compression cycle heat pump with a compressor
9. Open air cycle heat pump with a compressor
10. Chemical heat transformer
11. Ranque-effect-based heat pump
12. Electrodiffusion-effect-based heat pump
13. Thermoelectric heat pump
14. Magnetic effect heat pump

The diagram of a vapour compression cycle heat pump installation with a compressor and single-component working fluid is shown in Figure 1.4a. The working fluid vapour, due to be mechanically compressed (in thermodynamic state 1), flows into the compressor from the evaporator, which forms a connection with the low-temperature heat source. After compression, the vapour (now in state 2) flows into the condenser, which provides the heat pump with a connection to the high-temperature heat source. The vapour condenses at constant pressure. Following the Gibbs rule, this condensation also occurs at constant temperature. After condensation, the liquid (in state 3) undergoes expansion in an expansion valve. The expanded two-phase mixture (in state 4), most of which is a liquid, flows into the evaporator, where it also evaporates in constant-pessure, constant-temperature conditions, after which the vapour (in state 1) flows into the compressor, thus completing the working fluid flow cycle. The changes of the

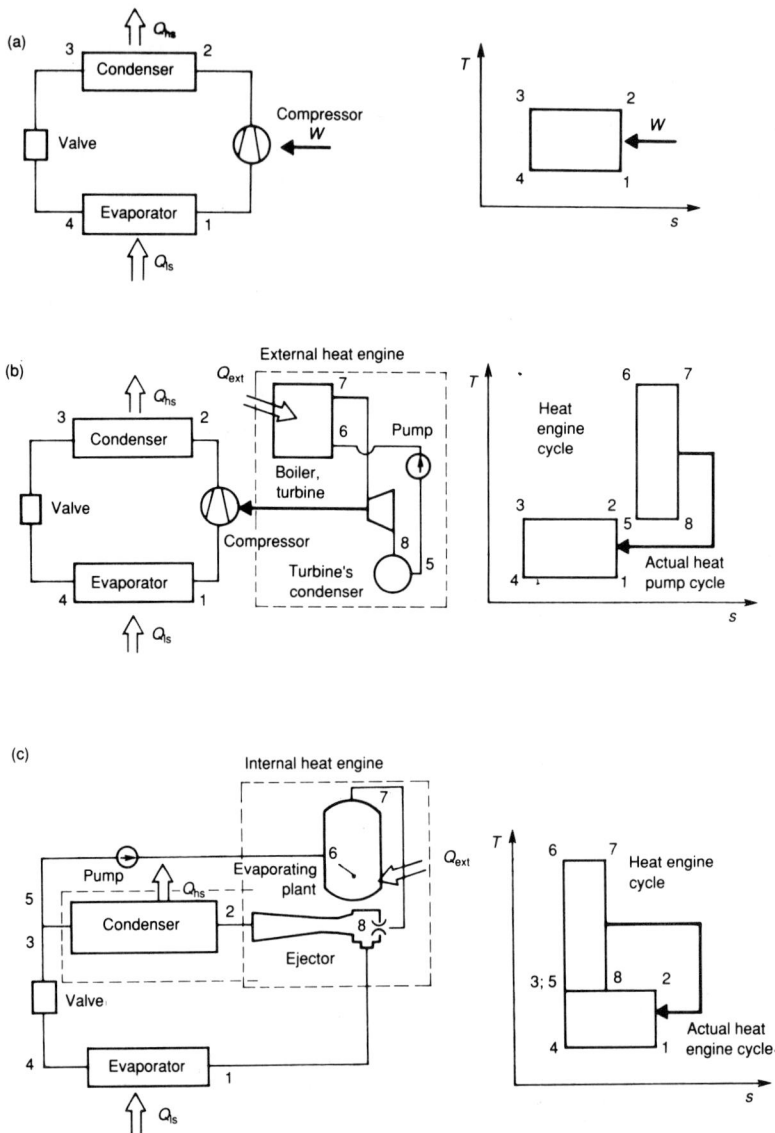

Figure 1.3   *Possibilities of using various forms of driving energy: (a) mechanical energy; (b) heat energy, converted into work in an external heat engine; (c) thermal energy, converted in an internal heat engine into the driving enthalpy, kinetic energy and enthalpy of the fluid; (d) thermal energy, converted in the internal heat engine – through the absorption and desorption processes – into the enthalpy of the fluid; (e) as above, but through the processes of condensation and evaporation*

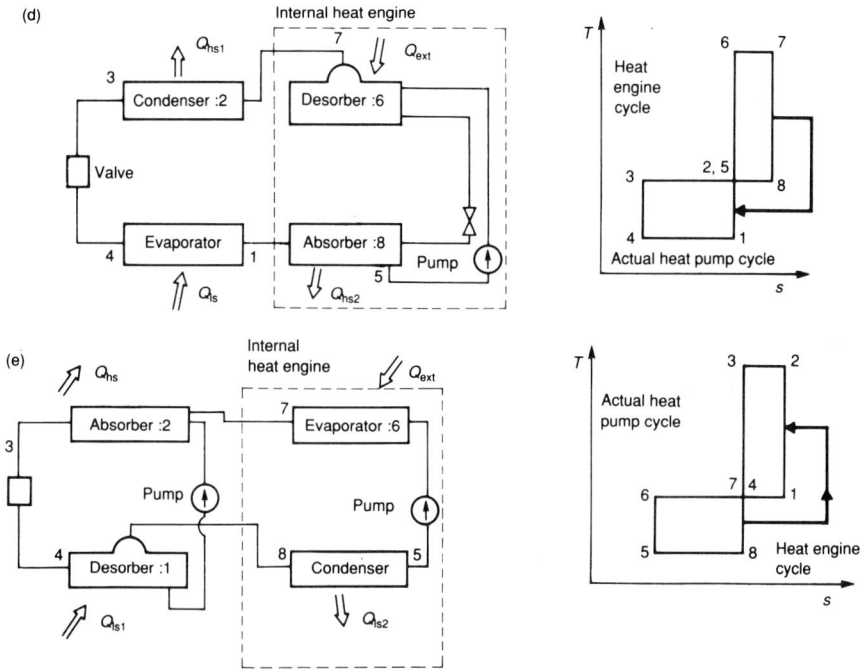

Figure 1.3   *(Continued)*

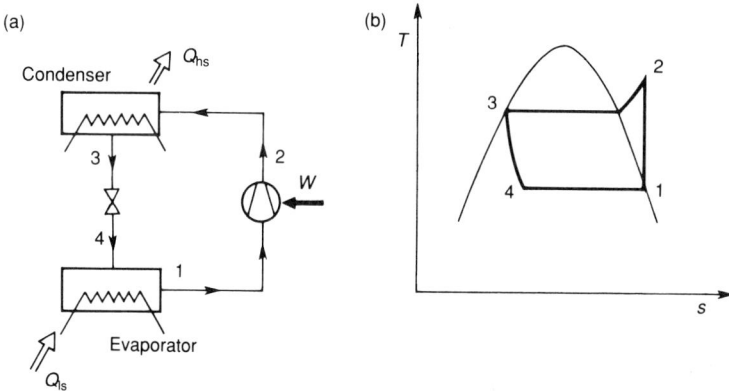

Figure 1.4   *Compressor-based heat pump with a single-component fluid:*
(a) schematic diagram with indication of salient point; (b) non-ideal reverse Rankine cycle

working fluid state in such an installation are illustrated by the theoretical non-ideal reverse Rankine (Linde) cycle (Figure 1.4b). Such a cycle describes these changes better than the Carnot cycle mentioned earlier.

The vapour compression cycle heat pump with a compressor and a two-component working fluid can be realized in the same installation as that described above (Figure 1.4a). The condensation and evaporation occur under constant-pressure conditions, but not at constant pressure and constant tempera-

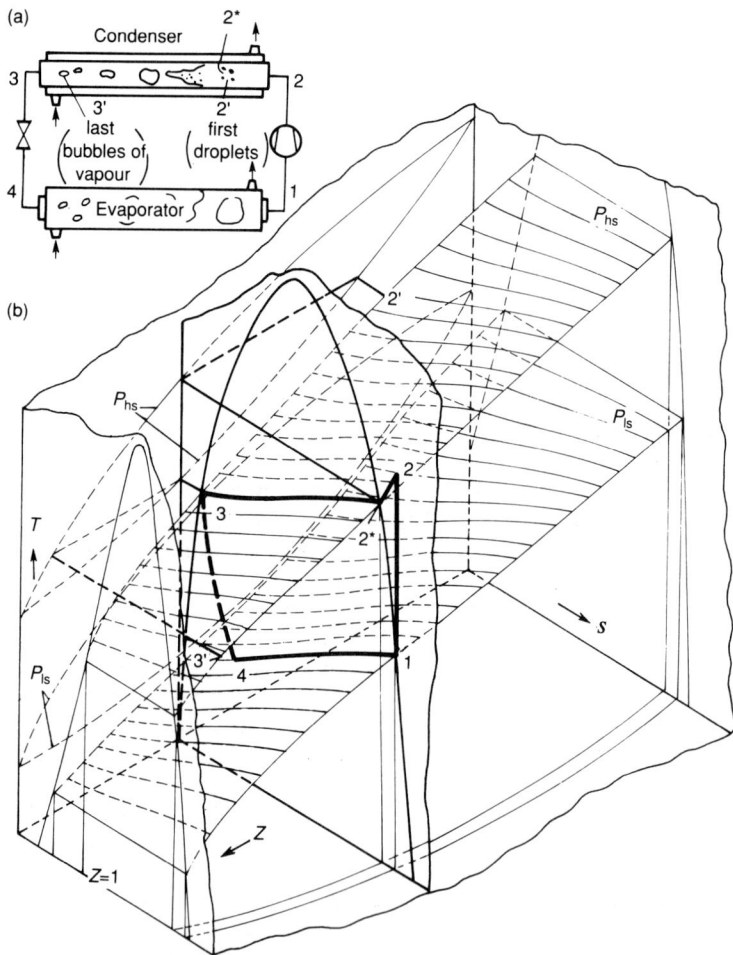

Figure 1.5 *Compression cycle, compressor-based heat pump with working fluid being a zeotrope mixture: (a) diagram of installation; (b) cycle representation in the T–s–z space; 2', state of the condensate at the condenser inlet; 3', state of vapour at the condenser outlet; (c) interpretation of this cycle in the T–s (z = 1) plane; (d) Lorenz cycle*

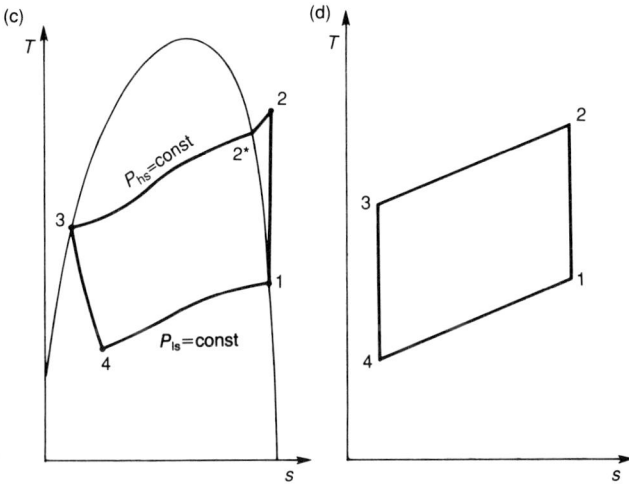

(c)

(d)

$P_{hs}=const$

$P_{ls}=const$

Figure 1.5   *(Continued)*

ture, as was the previous case. When the heat capacities of the sources (high and low temperature) are limited, such a cycle yields a higher exergetic efficiency of the condenser and evaporator, and thus of the heat pump as a whole. On top of this, such a cycle better reflects real conditions, where the heat capacities of sources are indeed limited, which leads to significant differences between the inlet and outlet temperatures of the source substances. This case is discussed in detail in Chapter 4; here we stop at illustrating the cycle in the three-dimensional $T–s–z$ space (Figure 1.5b) and presenting the ideal Lorenz cycle (Figure 1.5d), which may be used as a comparative cycle.

The cycle of a sorption heat pump, which is more complicated, can be interpreted in two ways. In the first version the working fluid undergoes the same cycle as in a compressor-type heat pump, but the vapour is compressed by heat supply. In the second version, the cycle of a sorption cycle heat pump really consists of two cycles: the heat engine cycle, which produces work, and the actual heat pump cycle, to which this work is transmitted.

The diagram of a sorption cycle heat pump installation is shown in Figure 1.6a, and the ways of accomplishing individual transitions are listed in Figure 1.2. The working fluid phase changes are realized in the same way as in a heat pump equipped with a compressor, and the working fluid vapours are compressed by sorption. The working fluid vapour, due to be compressed (in state 1), flows from the evaporator, which forms a connection with the low-temperature heat source of the actual heat pump cycle, to the absorber, which in turn is the low-temperature heat source of the heat engine cycle. In the absorber, in the presence of the sorbent, the vapour forms a two-phase and two-component mixture with the sorbent, at the same pressure but at a higher temperature (state 5′). The vapour then condenses, the working fluid is absorbed (state 5), and the liquid mixture of working fluid and sorbent on leaving the absorber is called the rich solution. This

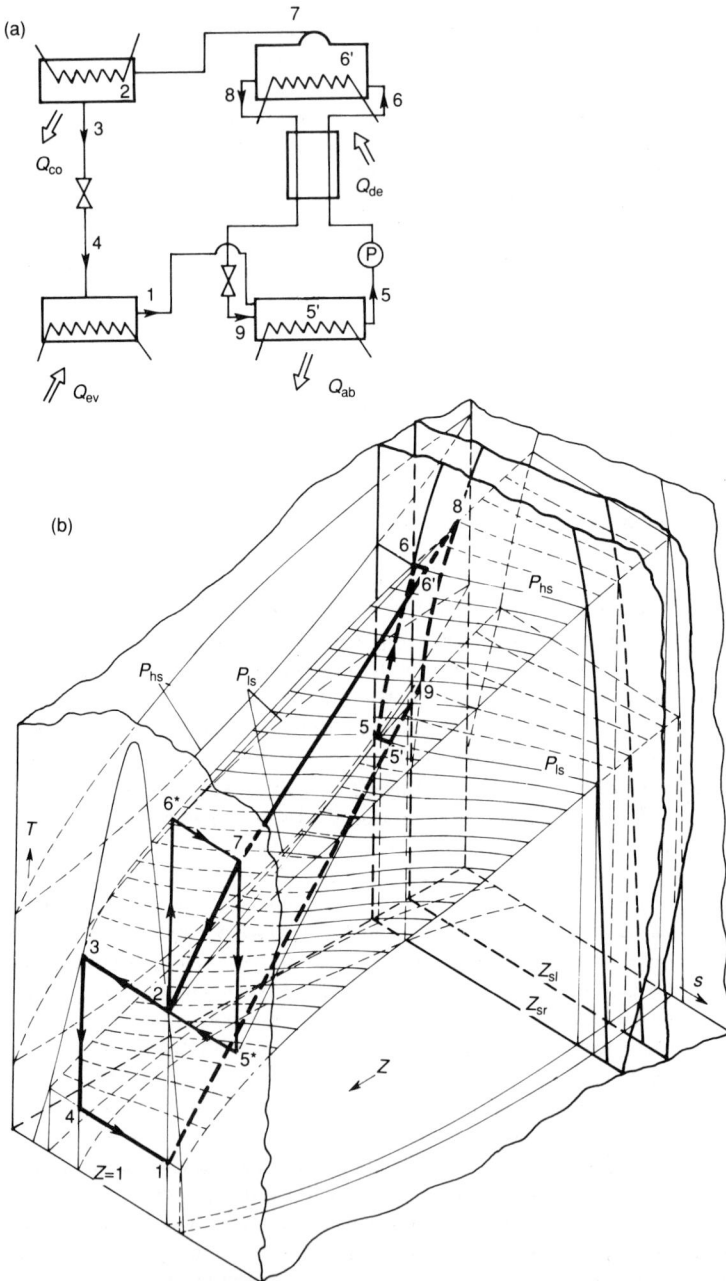

Figure 1.6   *Absorption cycle heat pump:* (a) *schematic diagram of installation;* (b) *cycle representation in the T–s–z space. The actual heat pump cycle is characterized by points 2341, and the heat engine cycle by points 26*75* obtained by projecting points 566'5' onto the plane T–s (z = 1)*

solution is next mechanically pumped to the desorber, which is connected to the high-temperature heat source of the heat engine cycle. In the desorber, this rich solution forms a two-phase two-component mixture (defined by state 6'). The working fluid vapour (in state 7), separated from this mixture by desorption (the vapour can be rectified to increase the fraction of the working fluid), enters the condenser, which is connected to the high-temperature heat source of the actual heat pump cycle. In the condenser this vapour is superheated relative to the prevailing pressure and is first cooled (to state 2), and then condenses (to state 3). After condensing, the liquid (in state 3) is expanded in the expansion valve. The expanded (to state 4) two-phase (single-component) mixture evaporates (to state 1), and then enters the absorber, thus completing the working fluid cycle.

The installation described above requires that the sorbent be continuously supplied to the absorber and removed from the desorber. In technical implementations the sorbent circulates between the absorber and desorber as a liquid mixture called the lean solution ($z = z_{sl}$). A regulating valve is used to provide the necessary pressure difference. Maintenance of the necessary temperature difference between absorber and desorber can, on the other hand, be ensured in two ways. The first requires using suitably large retention volumes of solutions, both in the absorber and in the desorber, and the second requires a heat exchanger. This heat exchanger is called a regenerative solution heat exchanger. Both systems are, however, often used simultaneously (when the retention volumes are small).

Thermodynamic states of the working fluid are shown in Figure 1.6b. In the installation under consideration, because of the ongoing processes of sorption, the fraction of working fluid is different in the mass flow between the desorber, condenser and evaporator than in the flow between the absorber and desorber. Moreover, the magnitudes of flows are different. Because of that, a pseudo-intensive parameter z is introduced to characterize this fraction. Therefore, the points shown in Figure 1.6b, strictly speaking, do not represent a classical thermodynamic cycle. As is commonly known, such a cycle should be referred to a constant quantity (1 kg) of the working fluid. Nevertheless, the points shown in the plane $z = 1$ characterize the phase changes of the fluid in the actual heat pump cycle. It is easy to see the analogy to the non-ideal reverse Rankine cycle (Figure 1.4b). The heat engine cycle, in which the supplied heat is converted into the work of compressing the working fluid vapour, is described, on the other hand, by points located in two planes: $z = z_{sl}$ and $z = z_{sr}$. Projection of these points onto a single plane, $z = 1$, yields (after Carnotization) two cycles: for the heat engine and for the heat pump (Figure 1.3d).

The diagram of a sorption cycle heat transformer installation is shown in Figure 1.7a, and the ways of accomplishing the individual transitions are listed in Figure 1.2. In contrast to the absorption cycle heat pump, described above, the changes of physical state are accomplished through sorption processes, and the fluid's vapour is compressed 'thermally'. The vapour, due to be compressed (in state 1), flows from the desorber, which now forms a connection with the low-temperature heat source of the actual heat pump cycle, to the condenser, which in turn is the low-temperature heat source of the heat engine cycle. In the condenser this vapour is superheated relative to the prevailing pressure and is first cooled (to state 5), and then condenses (to state 6). The condensate obtained in this way is next

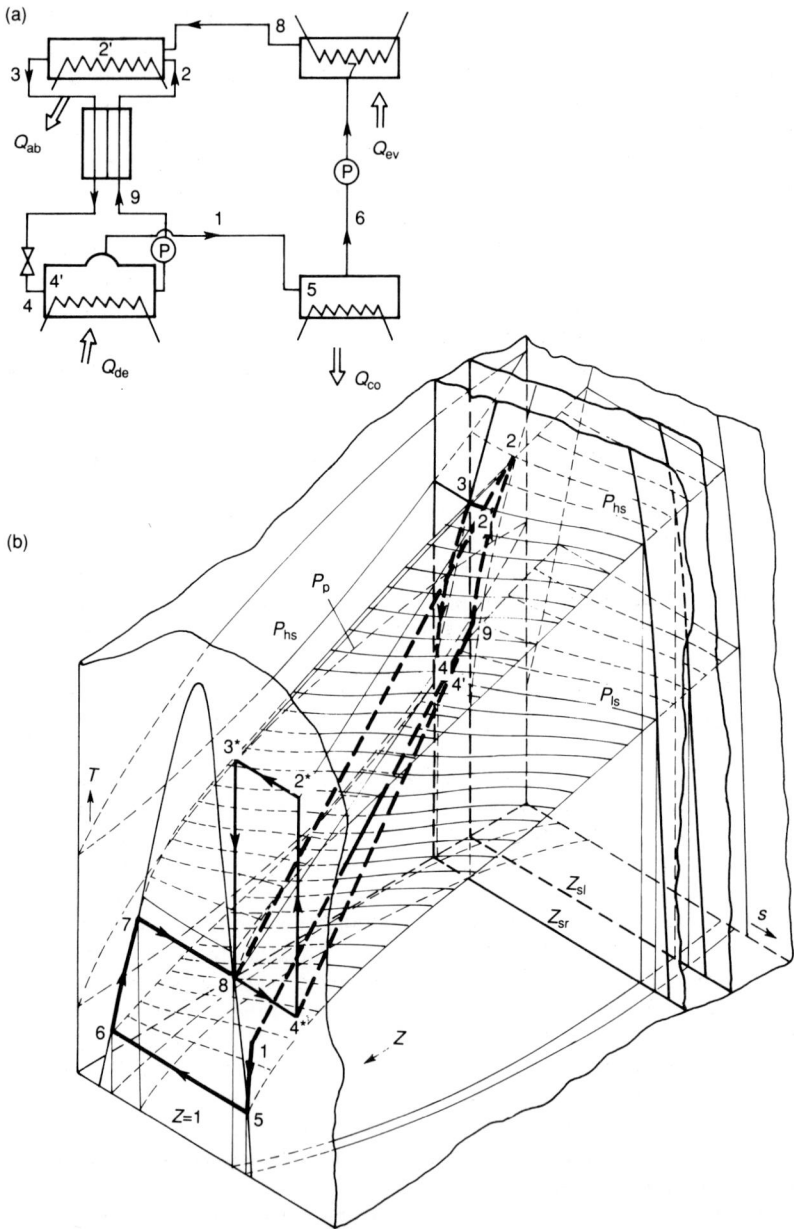

Figure 1.7   *Absorption cycle heat transformer:* (a) *schematic diagram;* (b) *cycle representation in the T–s–z space. The actual heat pump cycle is characterized by points 44'2'3, its projection onto the plane T–s (z = 1) by points 84*2*3*, and the heat engine cycle by points 5678*

mechanically pumped to the evaporator, which is connected to the high-temperature heat source of the heat engine cycle. In the evaporator, the supercooled fluid is first heated (to state 7), and then evaporates (to state 8). The vapour so formed flows into the absorber, which is connected to the high-temperature heat source of the actual heat pump cycle. Here, in the presence of the sorbent, the vapour forms a two-phase and two-component mixture (state 2'). The vapour then condenses and the working fluid is absorbed (state 3). The rich solution leaving the absorber, after passing through an expansion valve, enters the desorber, connected to the low-temperature heat source of the actual heat pump cycle. In the desorber, the solution forms a two-phase, two-component mixture (state 4'). The fluid's vapour, separated from this mixture by desorption (state 1), enters the condenser, thus completing the cycle.

In the technical implementation of such an installation, the sorbent circulates between the absorber and the desorber. To provide the necessary pressure difference between these exchangers, the condensate is mechanically pumped. The temperature difference between these exchangers is maintained in a similar manner as in the absorption cycle heat pump.

Thermodynamic states of the working fluid are presented in Figure 1.6b. The physical phase changes of the fluid are accomplished through sorption processes, and thus the salient points are located in two planes: $z = z_{sl}$ and $z = z_{sr}$. The heat engine cycle is presented in the plane $z = 1$. When all these point are projected onto a single plane, one obtains (after Carnotization) the heat engine and heat pump cycles from Figure 1.3e.

The installation diagram of a resorption cycle heat pump with a mechanical compressor is shown in Figure 1.8a, and the ways of accomplishing the individual transitions are listed in Figure 1.2. The changes of physical state of the working fluid in the low- and high-temperature heat sources are accomplished through sorption processes, and the vapour, as in a compressor based heat pump, is compressed mechanically. The vapour, due to be compressed (in state 1), enters the compressor from the desorber, which forms a connection with the low temperature heat source of the heat pump. After compression, the vapour (now in state 2) enters the absorber, connected to the high-temperature heat source. To emphasize the fact that this absorption takes place at a pressure higher than that of the desorption process, in the literature the absorber is also called a resorber. After absorption, the rich solution leaving the absorber (in state 3) passes through an expansion valve and enters the desorber. In the desorber, the fluid's vapour is separated from the two-phase, two-component mixture (in state 4'). The vapour (now in state 1) enters the compressor, thus completing the cycle.

Because the absorption and desorption processes are conducted in non-isothermal conditions, the COP of such a pump is higher than that of a compressor-based heat pump.

The diagram of a resorption cycle heat pump installation is shown in Figure 1.9a, and the corresponding thermodynamic states in Figure 1.9b. The salient points of the heat engine and heat pump cycles are located in different planes. Although pumps of this configuration are not used in practice, because of the greater COP the utilization of 'sorption' compression presents a solution which merits attention, particularly in some markets.

(a)

(b)

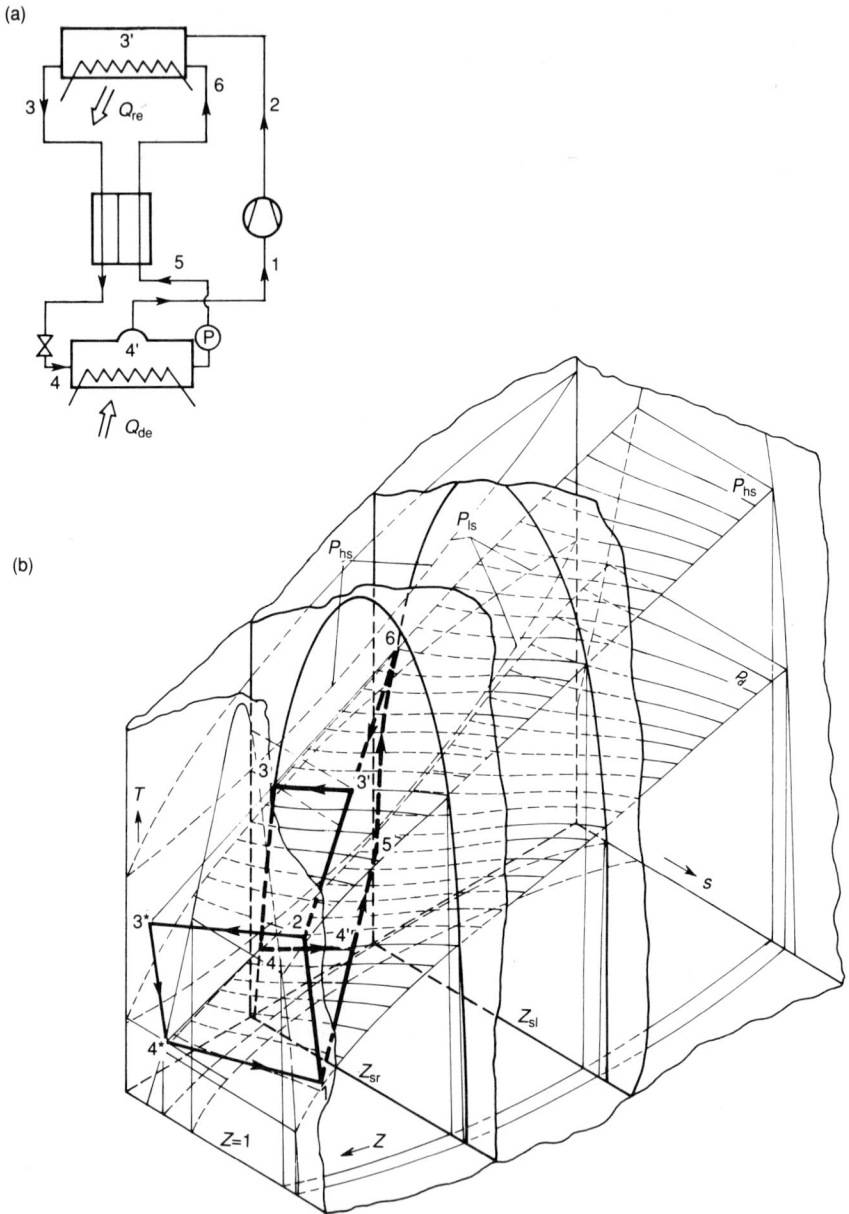

Figure 1.8   *Resorption cycle heat pump with a mechanical compressor:*
(a) *schematic diagram;* (b) *cycle representation in the T–s–z space. The actual heat pump cycle is characterized by points 3'344', and its projection onto the plane T–s (z = 1) by points 23*4*1*

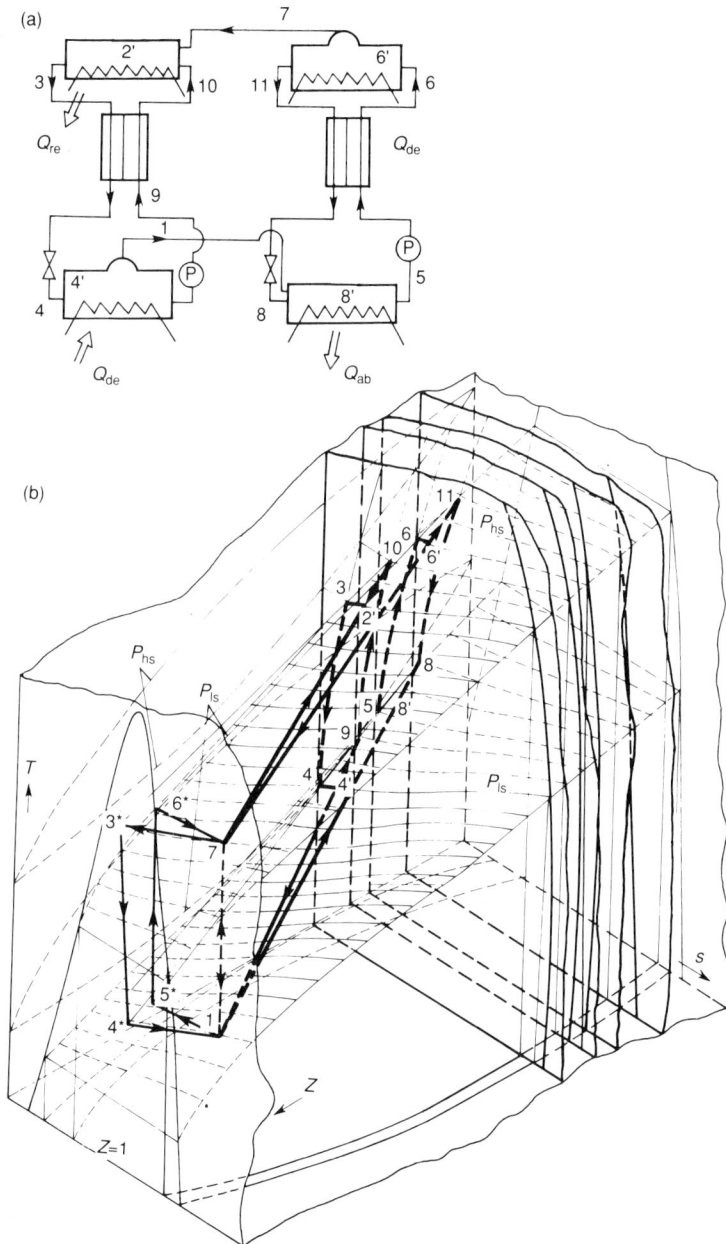

Figure 1.9   *Resorption cycle heat pump with sorption compression of the working fluid: (a) schematic diagram; (b) the actual heat pump cycle is characterized by points 2'344', and its projection onto the plane T–s (z = 1) by points 73*4*1; the heat engine cycle is characterized by points 8'566', and its projection onto the plane T–s (z = 1) by points 15*6*7*

The compression of vapours can be applied to various kinds of evaporation processes. In the majority of cases water vapour constitutes the vapour phase. The idea behind the compression of vapour is to raise its enthalpy with the aim of utilizing this enthalpy to produce further quantities of vapour. In a thermodynamic sense the compression of vapour thus corresponds to the implementation of an open heat pump cycle. The low-temperature heat source is provided by the boiling solution, and the high-temperature heat source by the condensing compressed vapour, which is at the same time heating the same solution. The working fluid of such a heat pump is the vapour itself.

In the food-processing industry, mechanical compressors or specially designed fans are being introduced, although compression with ejectors is more commonly used. These ejectors are driven by steam supplied from a boiler (live steam), and such a compression process is, not quite justifiably, called thermocompression.

The vapour compression process gives rise to an open cycle with a relatively small temperature difference, in the order of 10 K, and thus with relatively high

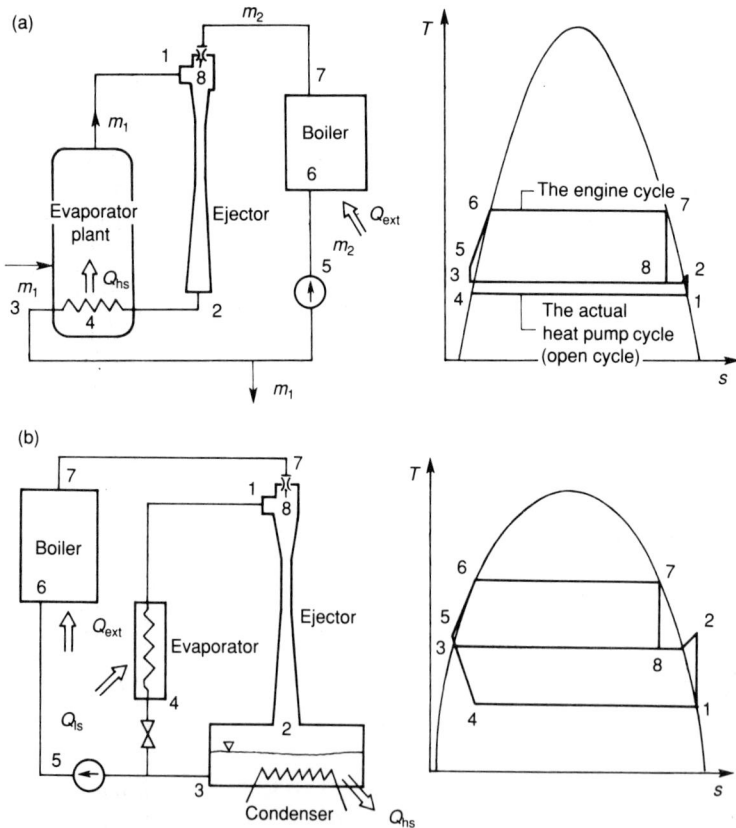

Figure 1.10   *Thermocompression of vapour:* (a) *schematic diagram of installation and the actual heat pump cycle;* (b) *ejector-based heat pump*

COPs (theoretically COP = 34 when $\Delta T = 10\,K$); hence the great attractiveness of this plant. Vapour-compression-based plant is not, however, discussed here in too much detail, for the following reasons: the compression of vapours does not differ much from the compression processes discussed in Chapter 4, and it is not easy to obtain the necessary compressors in Poland. Thermo-compression, on the other hand, introduces certain new elements, worthy of discussion, albeit also only briefly.

Thermocompression as a technical task is relatively well mastered. It is widely used for the vapour compression processes discussed here, as illustrated in Figure 1.10a. Thermocompression can be used to accomplish a cycle similar to that of mechanical compression, but containing an auxiliary heat engine cycle driving the ejectors. This type of heat pump was shown in Figure 1.3c. and its principle of operation is shown in Figure 1.10b. The working fluid may be the same as in the compressor based heat pumps. Because it now mixes in the ejector with the driving fluid (propellant), both the heat engine and the heat pump circuits must contain the same fluid. Details are presented in Example 6.3.

Figures 1.3c and 1.10 show two separated cycles: the heat engine cycle and the actual heat pump cycle. In Figure 1.3c the Carnot cycles are used to illustrate the fact that the engine work (area 5678) is the same as the work transferred to the heat pump. The use of intensive parameters allows, on the other hand, better illustration of the theoretical cycles (Figure 1.10).

A compressor based heat pump employing a gas cycle is shown in Figure 1.11. If an expander is included, this cycle can be compared to the theoretical Joule cycle. The COP of such cycles is not too big; hence the necessary use of the expander, which transfers the power to the compressor shaft. Without the expander, with just an expansion valve instead, the COP of such a cycle would be even lower. For this reason the gas cycle heat pump cannot compete with a compressor-based vapour cycle heat pump.

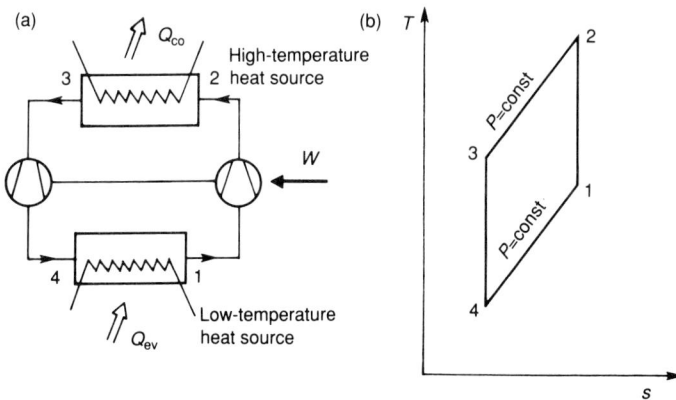

Figure 1.11 *Compression-type heat pump with gas cycle, mechanical compressor and expander:* (a) *schematic diagram;* (b) *comparative Joule cycle*

A compressor-based heat pump employing an open air cycle is shown in Figure 1.12. Such a pump can be used when the temperature differences between the high- and low-temperature heat sources do not exceed a dozen or so degrees, and the low-temperature heat source temperature is close to the temperature of the environment.

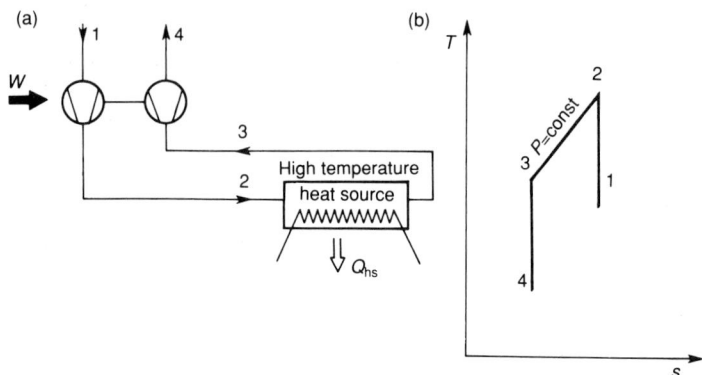

Figure 1.12   *Compression-type heat pump with an open air cycle, mechanical compressor and expander: (a) schematic diagram; (b) comparative Joule cycle*

The principle of operation of a chemical heat transformer is based on the utilization of thermal effects of endo- and exothermic chemical reactions. The utilization of heat effects of such reactions could be attractive because the volumetric heat flux (heat flow per unit volume) of generated heat is larger than for a vapour cycle. The detailed analyses of various possible chemical reactions, whose heat effects can be utilized in a heat pump, are given in [3], [4], [5], [6], [7], [8] and [9]. It proves difficult, however, to maintain over a period of time the constant properties of the fluid, as is the case with the physical processes of absorption and desorption. There is always a possibility that some secondary reactions can take place and yield trace quantities of other chemical compounds. These can accumulate in time, leading to serious changes in the original chemical composition of the working fluid, which in turn may lead to changes in the reactions taking place and their heat effects. One can also expect changes in the properties (activity) of the catalyst, which also has to reflect the changes in the type or intensity of reactions. For these reasons, heat pumps of this type do not leave the research laboratories, despite the interest in them, mentioned above.

The principle of operation of a chemical heat transformer is shown using the example of the endothermic dehydrogenation of 2-propanol to acetone, and the exothermic hydrogenation of acetone to 2-propanol. These reactions take place according to the following equations.

Endothermic dehydrogenation:

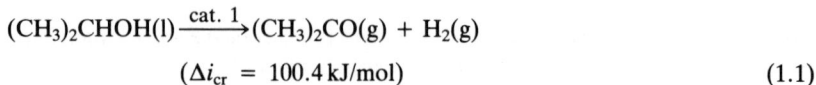

$$(CH_3)_2CHOH(l) \xrightarrow{\text{cat. 1}} (CH_3)_2CO(g) + H_2(g)$$

$$(\Delta i_{cr} = 100.4 \, \text{kJ/mol}) \tag{1.1}$$

Exothermic hydrogenation:

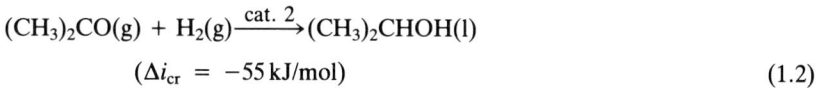

$$(CH_3)_2CO(g) + H_2(g) \xrightarrow{\text{cat. 2}} (CH_3)_2CHOH(l)$$

$$(\Delta i_{cr} = -55\,\text{kJ/mol}) \tag{1.2}$$

Heat absorbed by the endothermic reaction at the level of the low-temperature heat source is released at the high-temperature heat source level in the exothermic reaction. The work necessary to accomplish such tasks is obtained from the supplied energy (which is also supplied as heat). The above-discussed principle of operation is illustrated in Figure 1.13a, and Figure 1.13b shows a flow diagram of the heat and gas mixture flows. The diagram of installation is shown in Figure 1.13c.

Heat Q_{ext} is supplied at temperature T_{ext} to the endothermic reactor, which is designed as the boiler of the rectification column. Part of this heat is used to

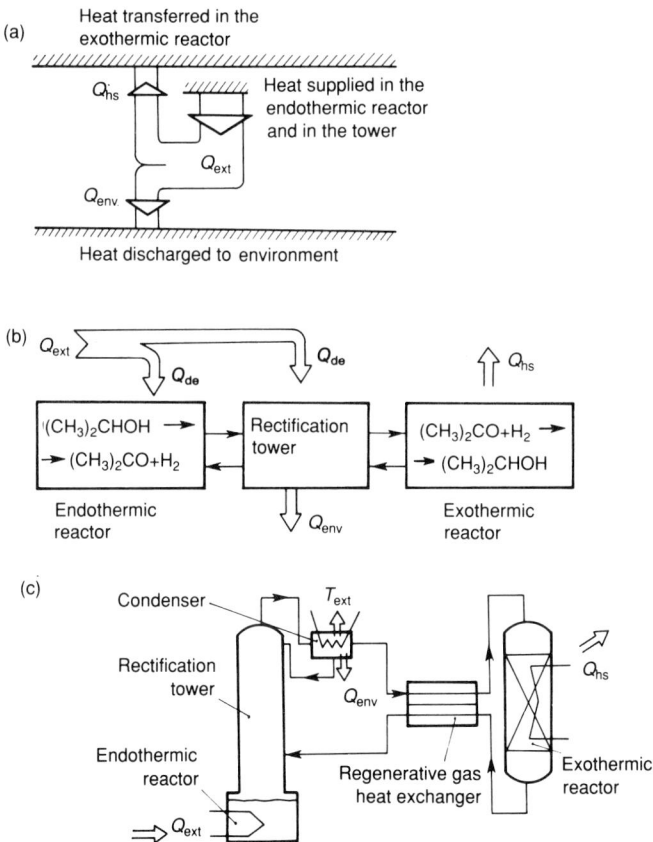

Figure 1.13 *Chemical heat transformer: (a) principle of operation; (b) diagram of mass and heat flows; (c) schematic diagram of installation*

maintain the reaction (Q_r), and part for evaporation (Q_{ev}). In the boiler a mixture of 2-propanol (CH_3)$_2$CHOH and acetone (CH_3)$_2$CO is boiling and simultaneously undergoing a hydrogenation reaction in the presence of a catalyst, at a pressure close to atmospheric. Acetone is the more volatile component, as at atmospheric pressure it boils at 329.3 K (2-propanol at 355.5 K). The acetone, together with hydrogen, is collected in gas phase at the top of the column and is directed to the exothermic reactor. Once there, it is hydrogenated to 2-propanol in a gas phase reaction on a catalyst bed, yielding at temperature T_{hs} the heat effect Q_{hs} equal to Q_r. Powdered nickel is used as catalyst for the endothermic reaction, and for the synthesis (hydrogenation) reaction the catalyst used is nickel on an activated carbon base. It can be added here that both 2-propanol and acetone fulfil the requirements set for the substances used in heat pumps (Section 2.1), although 2-propanol has recently been suspected to be a possible carcinogen.

Utilization of the Ranque effect in a heat pump is theoretically possible, but not economically viable. The effect itself is based on the existence of temperature differences in vortex flow. Ranque (1931) discovered this effect in a typical gas dust removal cyclone. However, it was only in the middle 1940s that Hilsh studied this phenomenon in more detail on a through-flow device (in the literature this device is referred to as the Hilsh tube; Figure 1.14).

The device separates the gas stream flowing tangentially into the tube (state 1) into two streams: hotter (state 2), which in a heat pump can be considered as the high-temperature heat source, and the cooler stream (state 3), which constitutes the low-temperature heat source. The separation into hotter and cooler streams, discussed here, occurs in both the subsonic and supersonic flows. The flow in a Hilsh tube proceeds as follows. The tangential direction of air inlet causes a vortex to form. Air is collected from both ends of the tube: hot air from the slot around the circumference, and cooler air from an opening located at the axis of symmetry. The temperature differences of the two air streams, generated during the flow through the Hilsh tube, can be caused by several effects: dissipation of kinetic energy (temperature increase), compression of air (temperature increase), expansion of air (temperature decrease), and finally centrifugal separation of cooler air of larger density (separation of cooler and hotter air). It is difficult, however, to predict the resultant effect of all the mentioned causes, particularly as the flow is not always stable. The tangentially fed input air stream causes swirling of different intensity in different parts of the tube. Due to this effect, pressure in the tube is also differentiated: it is higher at the walls and lower in the centre. Thus there can be expansion and the accompanying temperature drop along the tube axis, and temperature closer to the wall can be higher. On top of this, the dissipation of kinetic energy in the vicinity of the wall is larger due to friction and local losses, and hence the temperature increases can be larger. There is also a temperature increase close to the hot air outlet caused by the compression of air, resulting from the decrease of the average axial velocity (in this direction of flow). This velocity decrease results from the outflow of air along the tube axis towards the cold air outlet.

By attaching appropriate heat exchangers to the Hilsh tube one can put together a heat pump which implements the theoretical cycle shown in Figure 1.14c. In such a cycle, for $m_2/m_1 = 0.2$ and with the power necessary to drive the compressor

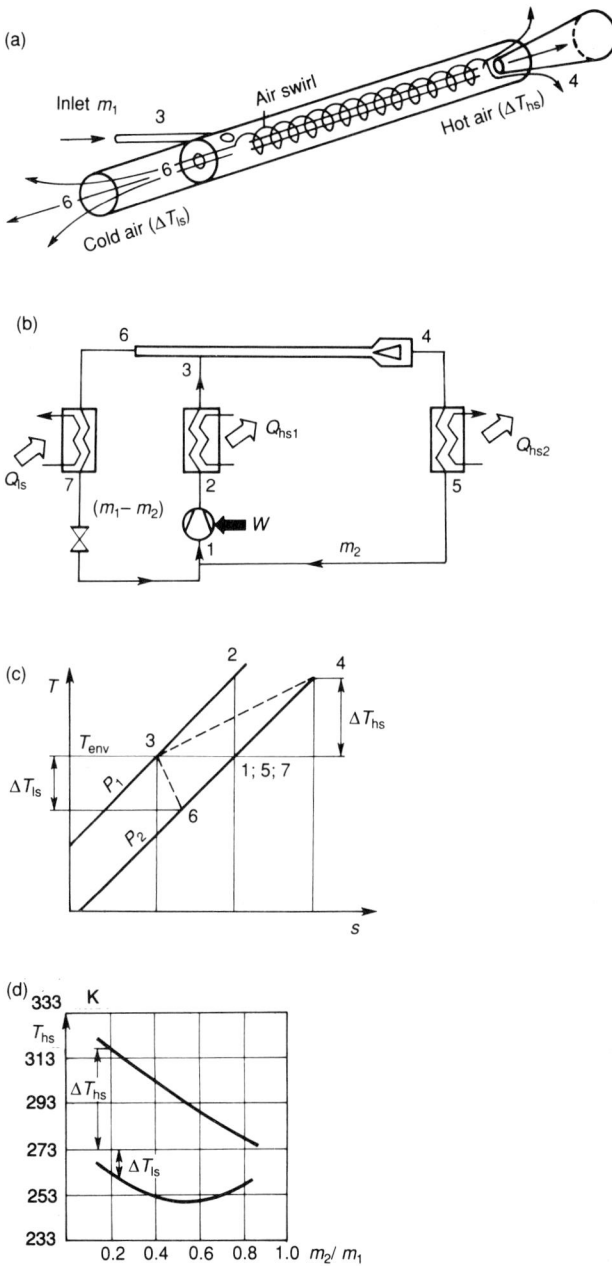

Figure 1.14 *Compression cycle heat pump based on the Ranque effect: (a) diagram of flows; (b) schematic diagram of installation; (c) representation of processes in the T–s plane; (d) results of investigations by Bonc et al. [10]*

$N = 5\,\text{kW}$, one obtains the heat power at the high-temperature heat source $Q'_{hs1} + Q'_{hs2} = 5 + 1 = 6\,\text{kW}$ with the heat flow absorbed from the environment $Q'_{ls} = 1\,\text{kW}$. The COP is thus 1.2 (in a compressor-based vapour compression cycle it is nearly three times higher). Taking into account the necessity of installing a compressor which has to be driven by mechanical energy, this pump cannot compete with a compressor-based heat pump. An additional drawback of this pump is its flow-induced noisiness.

Operation of an electrodiffusion effect heat pump is based on direct use of electrical energy for compressing the vapours of the working fluid. The electrical energy is supplied to a high-temperature resistant solid-state electrolytic material, which can conduct metallic ions. This material is called BASE (beta-aluminium solid electrolyte).

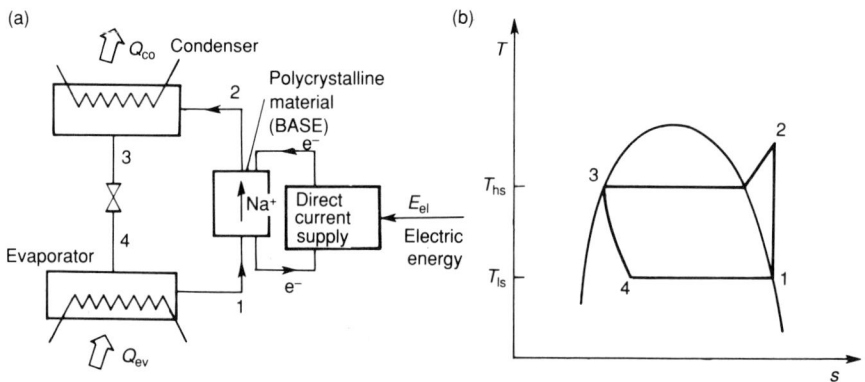

Figure 1.15 *Heat pump based on the electrodiffusion effect, using sodium as the working fluid:* (a) *schematic diagram of installation;* (b) *non-ideal reverse Rankine cycle*

The diagram of the heat pump is shown in Figure 1.15a. The liquid working fluid (sodium) evaporates in the evaporator and its vapour is next adsorbed by a porous anode, which forms a coat on the electrolyte supplied with DC current. In the electrolyte the sodium ions migrate through the polycrystallic BASE structure (the direction of flow of electrons is indicated on the diagram). After recombining with the electrons, the sodium vapour atoms have a higher electrochemical potential, and thus higher pressure and temperature. The sodium vapour condenses in the condenser, releasing heat at the high-temperature heat source. Thereafter, the liquid sodium is throttled in the expansion valve and flows into the evaporator.

The thermodynamic cycle is shown in Figure 1.15b, and Table 1.2 presents the results of calculations, according to [11], with the isentropic compression efficiency $\eta_{ic} = 0.9$. This efficiency depends on the quality of the electrolyte and on its geometric parameters (surface area and thickness of the layer), and also on the quality of transformation of alternate to direct current. For the temperature range

Table 1.2 *Thermodynamic parameters of the cycle of an electrodiffusion-effect-based heat pump*

Point number	Temperature (K)	Entropy (kJ/(kg K))	Enthalpy (kJ/kg)
1	950	4.6085	4360.5
2	1517	4.6085	5010.6
3	1082	4.0620	379.8
4	950	0.4183	379.8

as presented in Table 1.2, the COP, defined as the ratio of heat flow rate in the high-temperature heat source to the electrical power supplied, equals 6.51. Taking into account the efficiency of electricity generation, equal (in the USA) to 29.3%, the COP related to the primary energy equals 1.91. This approximate analysis did not take into account any exergy losses resulting from heat flow in the direction opposite to the flow of sodium ions, caused by the temperature differences in the electrolyte material. Solutions similar to the one described here were proposed by Alafeld[12],[13] and Argabright[14].

The described heat pump can potentially be used at high temperatures (which are impossible to obtain using a vapour cycle), e.g. to recover the waste heat released by the coal gasification process or by ethylene manufacturing plant.

The utilization of thermoelectric effects for a heat pump cycle is theoretically possible, but not done in practice. Although refrigerators utilizing these effects have been used since the introduction of semiconductors, the powers obtained are quite low (less than a kilowatt; only a few designs have power in the order of several hundred kilowatts). Both the p- and n-type semiconductors are used; these consist of alloys including tellurium, bismuth, antimony and trace amounts of selenium. Addition of mercury dichloride leads to the formation of the n-type semiconductor, and the addition of bismuth telluride leads to the p-type semiconductor. Thermoelectric effects, include the Seebeck effect (1822) and Peltier effect (1834). One can neglect the Thomson effect, which leads to the generation of heat in the circuit due to non-uniformities of temperature of the conductor, as it does not play any important part in the quantitative description (because of small changes of the α_{p-n} coefficient with temperature). The Seebeck effect describes a relationship between the thermoelectric voltage in the element and the temperature difference. For the conditions discussed below, the effect can be described by the following relationship: $E'_{el} = \alpha_{p-n} (T_{hs} - T_{ls})$, where E'_{el} is the thermoelectric voltage, and α_{p-n} is the Seebeck coefficient for the circuit containing the p- and n-type semiconductors. The Seebeck coefficient usually has a value of the order of $-200/\mu V/K$, while α_p has a value of the order of $+200\,\mu V/K$. The Peltier effect results in absorption of heat by the cold junction and rejection of heat by the hot junction during the flow of current. The heat flow rate is equal to $Q' = \pi_{p-n}I$, where the constant π_{p-n}, defined by Thomson, is equal to $\pi_{p-n} = \alpha_{p-n}T$, T being the junction temperature (depending on whether the description relates to the cold or hot junction, T will be correspondingly equal to T_{ls} or T_{hs}). Thus the Peltier-effect-generated heat flows through the thermoelement from the low- to the high-temperature heat source (heat flow rate Q'_P), so that a heat pump can be realized (Figure 1.16). The described process is reversible, but it is accompanied by three irreversible processes, causing heat losses: production of Joule's heat due to resistance of the element, heat flow from the low-temperature heat source due to thermal conduction (heat flow rate Q'_k), and heat exchange with the environment (usually neglected in the analysis). These losses are quite significant, which is illustrated, after [15], by the flow diagram shown in Figure 1.16b as well as by the plots of the element's COP (Figure 1.16c) and of the exergetic efficiency (Figure 1.16d).

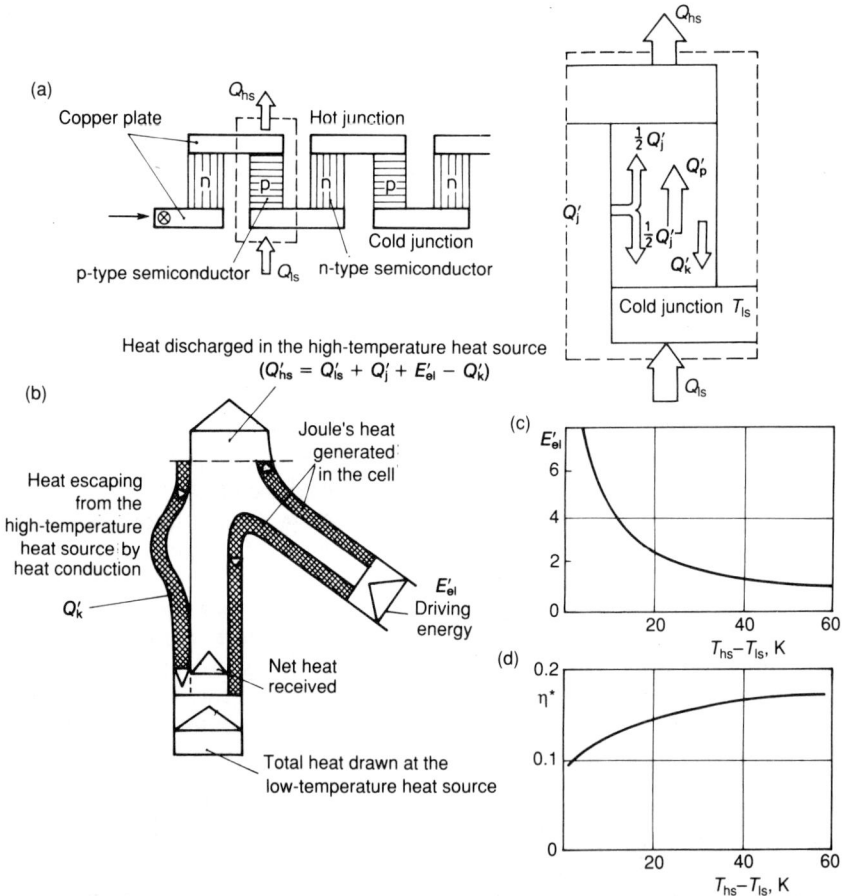

Figure 1.16   *Thermoelectric heat pump: (a) heat flow in the semiconductor element, (b) diagram of energy flow rates; (c) coefficient of performance of the thermoelement for $T_{ls} = 300\ K$; (d) exergetic efficiency of the thermoelement; $\rho_{el} = 800\ \Omega/cm$, $K = 15 \times 10^{-3}\ W/(m\ K)$*

The principle of operation of a magnetic heat pump is based on the magnetocaloric effect. The effect results in a change of temperature of the substance depending on the intensity of the magnetic field. The magnitude of this effect, measured by the temperature change, depends on the field intensity, which can reach a value of several, and in some modern designs of magnets even a dozen or so, teslas.

Figure 1.17a shows this effect for gadolinium (chemical symbol Gd, atomic mass 64, member of the lanthanide series). The presented data relate to a 50-g sample of cylindrical shape; temperature was measured in the cylinder's centre. It should be

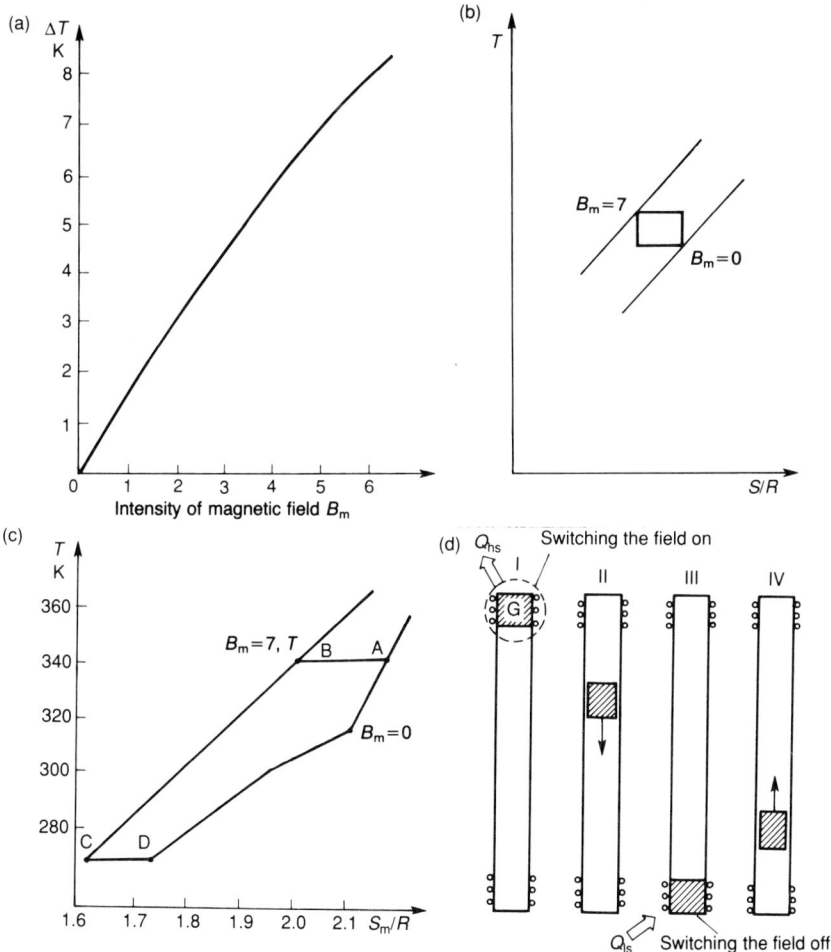

Figure 1.17   *Magnetic heat pump: (a) magnetocalorific effect for gadolinium; (b) Carnot cycle; (c) the cycle of NASA-built heat pump; (d) diagram of the technical implementation. I, heat transfer at the high-temperature heat source; II, regeneration (cooling) of the sample by the fluid; III, heat absorption at the low-temperature heat source; IV, regeneration (heating) of the sample by the fluid*

stressed here that gadolinium is characterized by the Curie point temperature (when its properties change from ferromagnetic to paramagnetic) of 293 K. At temperatures close to the Curie point the magnetocaloric effect has the largest magnitude. That is why gadolinium was selected by NASA as the material from which this heat pump was built.

It is postulated in the quantitative description of the magnetocaloric effect that the total entropy is a sum of the entropies of the crystalline lattice, free electrons and the magnetic field. This last entropy decreases with the increasing intensity of the magnetic field, due to its order-inducing effect on the spin orientations. Because of this, when the sample interreacts with the magnetic field under adiabatic conditions, the decrease of magnetic entropy is compensated by an increase in the entropies of the crystalline lattice and of the free electrons. This leads to the increase of the sample's temperature.

Figure 1.17b presents the total entropy change of gadolinium for two values of the field intensity. A Carnot heat pump cycle consisting of the isentropic–isothermal magnetization and demagnetization is also indicated in this figure. It has to be stressed that the temperature difference between the two sources in such a cycle is of the order of only a few Kelvins. To increase this temperature difference, and thus to increase the COP of the heat pump, a cycle shown in Figure 1.17c was realized. This cycle consists of two isotherms and two lines of constant strengths of magnetic field. Such a cycle can be approximated by the Ericson cycle. A heat pump implementing this cycle is shown in Figure 1.17d. The main element of this heat pump is a container with gadolinium, fabricated as a wire mesh to increase the heat transfer area. This container is located inside a cylindrical channel having diameter 50×10^{-2} m and length 1 m, filled with a mixture (50% water and 50% ethanol). The channel, together with the container, can move in a vertical direction and its ends are connected to the exchangers tapping the low- and high-temperature heat sources of the heat pump.

Description of the cycle can start with the isothermal magnetization (Figure 1.17d I). This magnetizing begins when the container with gadolinium is in contact with the high-temperature heat source (line AB in Figure 1.17c). Thereafter the container moves relative to the liquid contained in the cylinder, and its temperature changes (Figure 1.17d II, which corresponds to the line BC in Figure 1.17c). In the bottom location the magnet is switched off (Figure 1.17d III, which corresponds to line CD in Figure 1.17c). Heat has to be supplied to maintain the constant temperature of the sample (low-temperature heat source). The cycle is completed when the sample is again moved relative to the fluid all the way to the top location. A cyclic displacement of the sample through approximately 50 cycles eventually yielded low-temperature $T_{ls} = 243$ K and high-temperature $T_{hs} = 328$ K.

1.4.2 The coefficient of performance of a heat pump

The COP was introduced to facilitate thermodynamic assessment of a heat pump; it is also used for the economic analysis of its applicability:

$$\text{COP} = \frac{\text{(heat energy discharged at the high-temperature heat source level)}}{\text{(energy to drive the heat pump)}} \tag{1.3}$$

Such a definition of the COP does not, however, lead to an unequivocal assessment. This fact is particularly evident when different heat pumps are being compared, as the COP varies (see Table 1.1) from 0.5 to 4.5, and can even reach a value of 9. As is commonly known, the assessment of similar performance parameters of a heat engine (efficiency) or a refrigeration plant (the cooling effect) is also subject to some uncertainty, although in these cases this does not lead to such different interpretations as is the case for heat pumps. The problem with heat pumps is the possibility of different definitions of the driving and received energy.

The definition of driving energy depends mainly on what kind of energy is delivered to the heat pump. This can most clearly be illustrated in the case of a compressor-based heat pump, which receives mechanical work. This work can be produced from the chemical energy of fuel in the heat pump location, or it can be obtained from electrical energy drawn from the electrical mains, also produced from the chemical energy, but at a location different to that of the heat pump, e.g. in a power station. Thus a problem arises: which of these energies should be taken as the driving energy. In other words, it is a question of which control surface was taken into consideration, so that the driving energy can be defined in different ways.

The definition of energy received at the high-temperature heat source level can also be equivocal. The total energy flow rate there comprises two streams: energy transferred at the high temperature heat source of the actual heat pump cycle and also energy obtained from recovering (by heat regeneration) the losses which are inevitable during the conversion of driving energy into mechanical work. When this conversion is not a complex one, e.g. when mechanical work is obtained from electrical energy, the losses are small, and thus not much energy can be recovered by regeneration. In this case most of the heat energy originates from energy received at the high-temperature heat source of the heat pump. However, when chemical energy is used, the conversion process is much more complex and the losses are correspondingly larger.

1.5 Auxiliary concepts and definitions

1.5.1 Thermodynamic cycles used in heat pump description

The concept of a thermodynamic cycle is commonly known, yet its interpretation can differ, so that the concept itself is worth discussing, if only to avoid ambiguity.

Generally speaking, a thermodynamic cycle is a set of processes, after which the state of a working fluid returns to the starting point; this cycle is reproduced identically during consecutive time intervals in an installation containing the circulating working fluid. The very concept of a cycle originates from an idea of the fluid passing through the same thermodynamic states and returning to the starting state in each cycle. However, a situation is possible in which the consecutive processes are each applied to fresh fluid, which leaves the installation without attaining the starting state. One then refers to an open cycle, to underline the fact that the cycle is not completed, and the fluid at the outlet does not attain the inlet conditions.

For the purpose of a qualitative description (to describe a process realized in an installation), the concept of a cycle is used in four different forms with different meanings. The first of these largely identifies the concept of a cycle with the principle of operation of a heat pump. Hence the very word cycle, supplemented by the necessary information on the method of implementation of the process, to a large extent defines the principle of operation of a particular heat pump. Thus we have such cycles as: compressor-based vapour compression heat pump, absorption heat pump, absorption heat transformer, resorption heat pump with sorption compression, etc. The second form results from the fact that in heat pumps driven by heat energy we encounter a double cycle, namely the heat engine cycle and the actual heat pump cycle. The word cycle refers to the double cycle as well as to each of the constituent cycles. We can thus say that, for example, the cycle of an ejector heat pump (with thermocompression) consists the engine cycle and the actual heat pump cycle. The third and fourth forms are of only supplementary character. The third form relates to the kind of fluid employed, so that we can have a gas cycle when the working fluid is a gas, or a vapour cycle when the working fluid forms a two-phase system: liquid–vapour. The vapour cycle is so common in the sorption- and compressor-type heat pumps that the word is not even always mentioned. The fourth form relates to the direction of work interacting with the cycle. When work is supplied to the cycle, as is the case in heat pumps, then the cycle is composed from a sequence of processes proceeding (for example in T–s coordinates) in the anticlockwise direction. We then refer to an anticlockwise (or left-hand) cycle. The opposite case is that of the engine cycle, from which work is extracted, and the processes follow each other in a sequence which can be referred to as a clockwise (right-hand) cycle.

For the purpose of quantitative analysis of heat pumps, various kinds of comparative cycles are used. These comparative cycles can be calculated theoretically, without the need for experiments. There are many such cycles, approximating with varying degrees of accuracy the cycles actually realized in heat pumps. The Carnot cycle is quite fundamental for the description of heat pumps. It is an ideal cycle, i.e. all its processes are reversible. Among other ideal cycles are the Ericson, Stirling and Joule cycles, illustrated in Figure 1.18. The advantage of

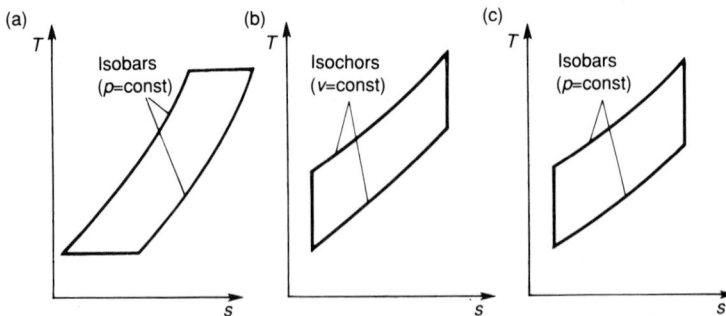

Figure 1.18 *Various ideal cycles:* (a) *Ericson;* (b) *Stirling;* (c) *Joule*

the Carnot cycle is that it can be described using only two intensive parameters, T_{hs} and T_{ls}, which is the reason it is so widely used as a comparative cycle. They do not, however, adequately describe the cycles actually realized on heat pumps. Because of this, other theoretical cycles are defined, often containing one or more irreversible processes. Thus one can talk about, for example, a theoretical cycle of a compressor-based heat pump, but its full character is defined by a detailed name.

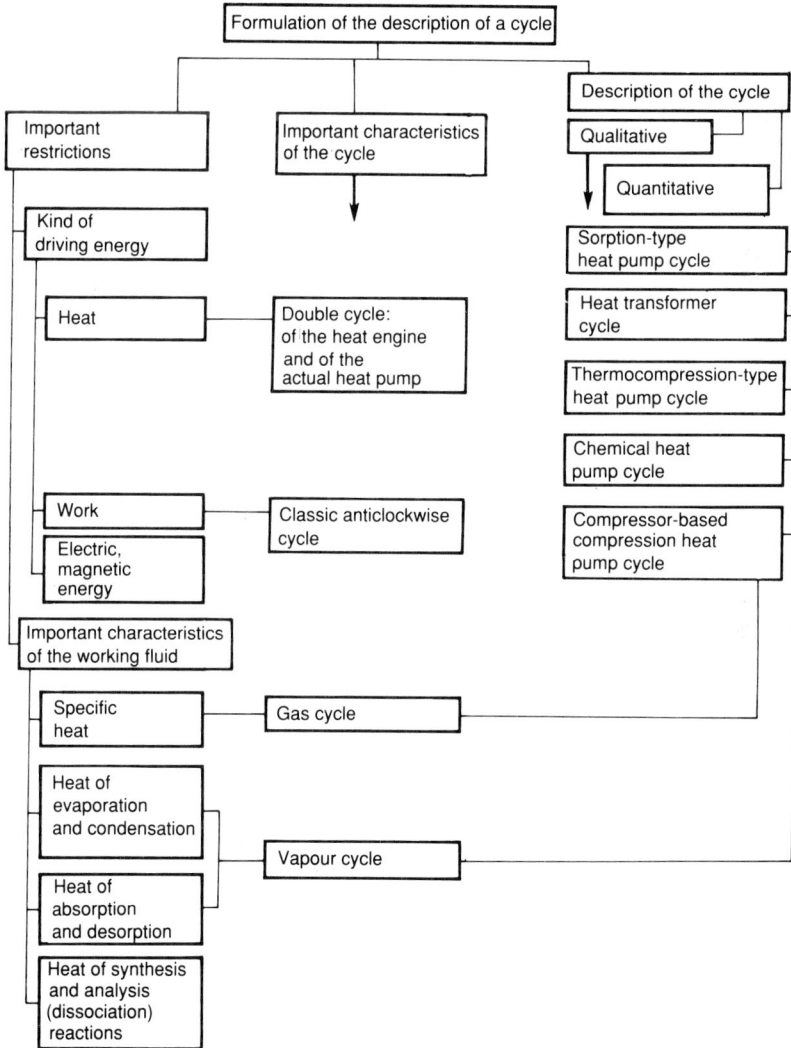

Figure 1.19 *Conditions and relations leading to the definition and description of a cycle*

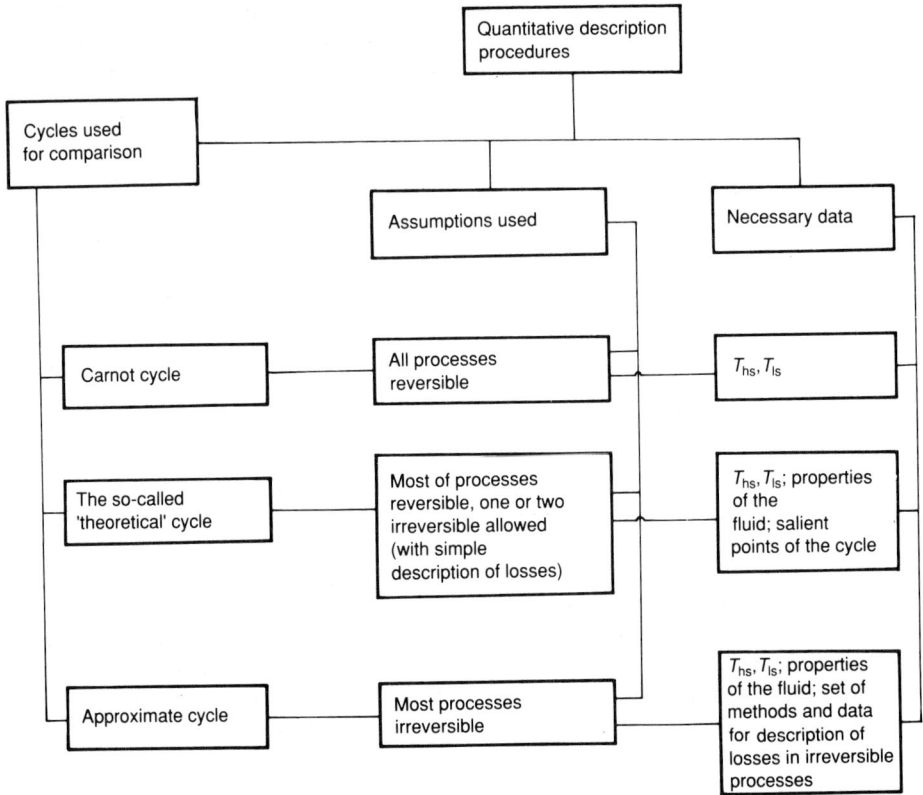

Figure 1.20   *Possibilities of description on various levels, starting from the simplest (Carnot cycle), to one that is more complicated and thus closer to a real cycle*

This is the case for the non-ideal reverse Rankine cycle, which unequivocally defines thermodynamic states of the fluid; the fluid is isentropically compressed – the state of liquid after condensation is defined by the boiling line, that of the vapour – by the saturated vapour line. One can also talk about a theoretical cycle of a sorption heat pump with isobaric–isothermic sorption processes, irreversible throttling processes, irreversible heat transfer processes and the related losses in regenerative heat exchangers. These losses are discussed further in the following chapters.

The description of theoretical cycles is not as simple as that of the Carnot cycle. It requires a detailed knowledge of the properties of the fluid and also requires taking into account the above-mentioned qualitative assumptions relating to the processes and states of the fluid in some point of the cycle. Also necessary for their description are the intensive parameters in the form of T_{hs} and T_{ls}.

Another tool in the description of the actually realized cycle is provided by the approximate cycle concept. This is also a theoretical cycle, but containing all the irreversible processes which result in losses. These losses are now an integral part of the description of a cycle, and since full unambiguity in their description is difficult to achieve, one can question whether such an approximate cycle can really be considered theoretical. The approximate cycle is treated as a next iteration in the theoretical description; for example, in case of a compressor-based (vapour compression cycle) heat pump, the first iteration is provided by the Carnot cycle, the second by the non-ideal reverse Rankine cycle, third by the non-ideal reverse Rankine cycle, and the fourth by the approximate cycle of a vapour compression heat pump with compressor (Figure 4.1). As the description of the losses may be perfected in various ways, the approximate cycle might be developed into consecutive iterations. The ambiguity in the description of losses is to some extent a weakness of the approximate cycle.

The interrelationships between the above discussed concepts, all of which use the word 'cycle', are presented in Figures 1.19 and 1.20, explaining the correlations between various cycles.

1.5.2 Characteristics of irreversible processes

It is known that the irreversibility of processes of transferring various kinds of energy results in the existence of driving forces, which can have the form of finite differences or gradients (for the heat transfer or mixing it is ΔT, for the flow of substance ΔP, for mass transfer Δz, for the flow of electric current ΔU). With the exception of heat energy, all other kinds of energy (and work) are partially dissipated and converted into heat energy, causing the increase of entropy. Thus the increase of entropy can be used as a measure of irreversibility of a process. This is illustrated here using an example of a heat exchanger.

The rates of flow of the heat supplied Q_1' and recovered (received) Q_2' (following the thermodynamics conventions) are defined as follows:

$$-Q_1' = \int_{S_1'}^{S_0'} T_1 \, \Delta S' \tag{1.4}$$

$$Q_2' = \int_{S_0'}^{S_2'} T_2 \, \Delta S' \tag{1.5}$$

The changes of entropy of the fluids transferring the heat are correspondingly equal:

$$S_0' - S_1' = -\Delta S_1' \tag{1.6}$$
$$S_2' - S_0' = -\Delta S_2' \tag{1.7}$$

It is known that the exchange process can only be realized irreversibly, i.e. with a finite temperature difference $T_1 - T_2 = \Delta T$. When $\Delta T = 0$, then $S_1' \rightarrow S_2'$ and thus

$$-\Delta S_1' + \Delta S_2' \rightarrow 0 \tag{1.8}$$

so there are no entropy changes. But when $\Delta T \neq 0$, then

$$-\Delta S_1' + \Delta S_2' = \delta S' > 0 \tag{1.9}$$

which is illustrated in Figure 1.21. It is now easy to prove the following relationship:

$$\overline{\Delta T}|\Delta S'| = \overline{T}\delta S' \tag{1.10}$$

where the bars above variables denote average values within the variation range. The quantity $\delta S'$ denotes increase of entropy. It is also a measure of the irreversibility of the process and is proportional to $\overline{\Delta T}$. It is easy to generalize this argument to other forms of energy, as shown in Table 1.3.

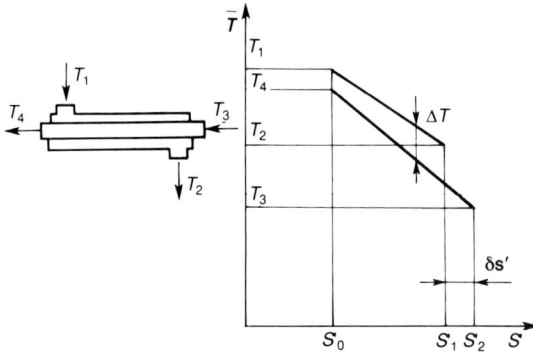

Figure 1.21   *Interpretation of the entropy production as a result of an irreversible heat transfer process*

Table 1.3   *Various forms of energy and losses resulting from their transfer*

Form of energy	Quantity	Losses as a measure of the process irreversibility	Increase of entropy
Thermal	$\int_{\Delta S} T\,dS$	$\overline{\Delta T}\,\Delta S = \overline{T}\delta\,S_{\Delta T}$	$\delta S_{\Delta T}$
Volumetric deformation (elastic)	$\int_{\Delta V} P\,dV$	$\overline{\Delta P}\,\Delta V = \overline{T}\delta\,S_{\Delta P}$	$\delta S_{\Delta P}$
Electrical	$\int_{\Delta U} Q_{el}\,dU$	$\overline{\Delta Q_{el}}\,\Delta U = \overline{T}\delta\,S_{\Delta U}$	$\delta S_{\Delta U}$

Where Q_{el} is an electric charge

It is not difficult to list the causes of individual irreversibilities, but it becomes possible only at the stage when detailed descriptions of the kind of heat pump involved and of the working fluid are available. One can then without too much difficulty indicate where and when irreversible processes and the acccompanying entropy increases will occur. First of all, we have here the heat or heat and mass transfer processes in exchangers, throttling of the working fluid in the expansion valve, and then for example, depending on the kind of heat pump involved, the compression process with accompanying throttling in the compressor valves, and also heat transfer between the cylinder walls and the fluid. In the case of heat pumps which have additional fans or circulation pumps installed, one has to remember that all the mechanical work used to drive them is also dissipated to heat.

The final effect of irreversibility of processes is the increase in the requirement for driving energy (to be more precise, exergy) to be supplied to the heat pump. It is well known that in the case of the heat engine, these irreversibilities result in a decrease of work obtained. The problem is illustrated here with an example of a compressor-based heat pump where the driving energy, supplied as work, is equal to exergy. In the case of other heat pumps the problem is similar and also results in an increase of the driving energy (thermal, chemical). In case of a compressor heat pump the minimum work requirement occurs for the Carnot cycle (Figure 1.22a).

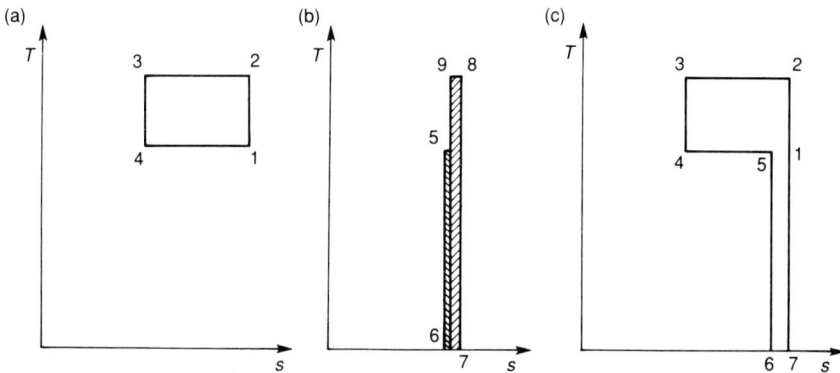

Figure 1.22   *Interpretation of the increase of work required in a cycle with losses: (a) work of the Carnot cycle (1234); (b) lost work (567895); (c) work of the cycle with losses (12345671)*

In a real (realized) cycle this work requirement is two to three times higher, which is a result of dissipation of work at the two hypothetical temperature levels T_{ls} and T_{hs} (Figure 1.22b). Figure 1.22c shows a real cycle having the same useful effect, i.e. the same amount of heat delivered at the high-temperature heat source, as in Figure 1.22a. The work delivered to drive the compressor is equal to the sum of work of the cycle and work dissipated to heat.

1.5.3 Exergy and exergetic efficiency

An important characteristic of every kind of energy is its relationship to the external conditions of the environment, particularly when conversions of energy from one kind into another are involved. In given conditions only a certain part of the energy can usually be utilized: only such part of the chemical energy which can be obtained from a reaction with the surrounding atmosphere at some specified conditions of pressure, temperature and concentrations of oxygen, nitrogen and (water) vapour, only such part of the potential energy which is defined by the displacement relative to the local datum; only such surplus of thermal energy which exists at temperatures higher than the temperature of the environment; and only such part of the elastic energy of volumetric deformation which exists at pressures higher than atmospheric. Since in heat pumps the problems of transferring the thermal energy are of particular importance, a few words follow on the definition of the surplus.

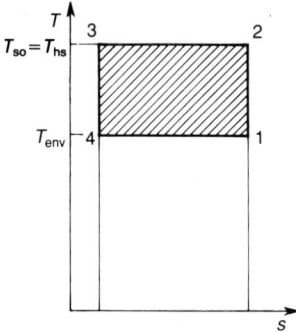

Figure 1.23   *Interpretation of exergy (1234) and thermal energy*

The surplus of thermal energy Q, transferred isothermally from the heat source of temperature T_{so}, called the exergy, is defined by a relationship resulting directly from Figure 1.23:

$$B = -\frac{T_{so} - T_{en}}{T_{so}} Q \tag{1.11}$$

where, in compliance with [16], the minus sign results from the fact that exergy is transmitted from the source. It can be seen from this relationship that exergy is decreasing with the decrease in temperature difference between the source and environment $(T_{so} - T_{en})$, which in some sense defines its degradation, meaning the loss of usefulness, when T_{so} approaches the temperature of the environment. This agrees with practice and is evidence of the usefulness of thermal energy.

An important part of the above definition of the surplus heat energy (or exergy) was the specification of the way in which the exchange with the environment proceeded (isothermally). The internal energy of a thermodynamic fluid can be converted to heat or work. Only a part of this energy (heat energy or work) relative to a certain level can be utilized and is called the maximum work. This concept is explained in more detail in [17] and [18]. The maximum work is a function of thermodynamic state, as is internal energy or enthalpy. For this reason its magnitude does not depend on the path of the process, but only on the initial and final states of the fluid. Such maximum work, whose final state is specified by the conditions of the environment, defines the exergy more generally than Equation

1.11, which only considers conversion of energy into heat. The maximum work takes into account the conversion of energy both into heat and work.

Exergy is not subject to the laws of conservation and the processes of exergy conversion are accompanied by its losses. These are related to the irreversibilities in the conversion processes. The exergy losses caused by irreversible phenomena within the system are described, according to the Gouy–Stodola law, by the relationship:

$$\delta B = T_{en} \sum_{i=1}^{n} \Delta S_i = T_{en} \delta S$$

where the summation sign denotes the increase in entropy of all the bodies participating in the phenomenon under consideration.

The concept of exergetic efficiency has been introduced in order to assess the degree of irreversibility of processes and the exergy losses. Exergetic efficiency can be described by two definitions, each of these comparing the real process (with its exergy losses) with an ideal process (in which there are no exergy losses).

The first definition, valid for the same useful effect, i.e. the same exergy transmitted in the source for both cycles, is as follows:

$$\text{exergetic efficiency} = \frac{\text{driving exergy of an arbitrary ideal cycle}}{\substack{\text{driving exergy of the real cycle, transferring in its} \\ \text{high-temperature heat source the same exergy as} \\ \text{the ideal cycle}}} \quad (1.12)$$

or:

$$\text{exergetic efficiency} = \frac{\substack{\text{exergy transferred in the high temperature heat} \\ \text{source of the real cycle}}}{\substack{\text{exergy transferred in the high-temperature heat} \\ \text{source of an ideal cycle driven by the same} \\ \text{exergy as the real cycle}}} \quad (1.13)$$

The concepts discussed here are used in the following chapters, when the energy conversions in heat pumps are described quantitatively. It is, however, worth stressing that if large exergy losses occur in an appliance (heat pump), its usefulness is greatly reduced. This is particularly important when other appliances can be used to perform the same function. Such is the case for a heat pump which has to compete with other installations (coal- or gas-fired heating, air conditioning), and these not only are able to deliver the energy at higher temperatures, but can more easily adjust to the required changes, both of temperature and of the heating power. These appliances are also subject to exergy losses, but since the temperatures at which heat is transferred are higher, these losses are not as significant as in a heat pump. The exergy losses in various domestic central heating systems, according to [19], are illustrated using flow diagrams in Figure 1.24.

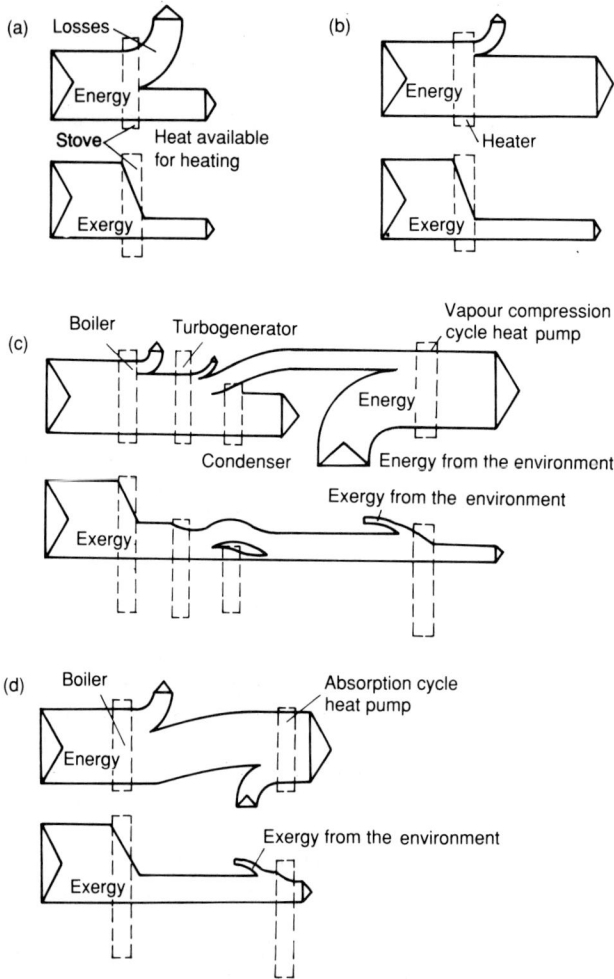

Figure 1.24   *Flow diagrams of energy and exergy for various heating systems:* (a) *stove;* (b) *electric heater;* (c) *vapour compression cycle heat pump;* (d) *absorption cycle heat pump*

The competitiveness of heat pumps in comparison with other heating systems must be eventually decided by economic conditions. For example, aiming at a maximum exergetic efficiency causes an increase in capital investment costs. The thermodynamic analysis can only deliver some of the indicators, although very important ones, necessary for this assessment. The exergetic analysis of various heating systems is given by Borel [20].

Chapter 2

Working fluids for vapour cycles, sorbents and working fluid–sorbent pairs

2.1 Working fluids

The kind of working fluid used is closely associated with the implementation of the cycle and with the installation used for the purpose. In the vapour compression and sorption heat pump cycles, the changes of phase of the working fluid (from vapour to liquid and vice versa) play an important part. Additionally, for the sorption cycle heat pump, the choice of a sorbent performing the role of a carrier of the working fluid, and the choice of a working fluid–sorbent pair are also important.

A working fluid is subject to several demands, which can be divided into general and particular ones. The most important general demands are market availability and the price, which should be relatively low. The working fluid should also display low aggressiveness towards the commonly used materials for plant elements, seals, measurement sensors, etc. Other desired properties are: non-toxicity, inflammability, stability, no odour; particularly for small heat pumps used in domestic heating.

The fluids used in heat pumps are selected from working fluids used in refrigeration technology, where the list of possibilities is quite long, reaching several hundred items. It should be stressed, however, that refrigeration fluids are not best suited to the demands of heat pump use, because of a different range of temperatures and thus also pressures.

The pressure increase is not high enough to cause any significant technical difficulties, but it causes a related increase in the installation costs, which reduces the attractiveness of heat pump use. Refrigeration does not have competing technologies of cooling, but a heat pump must compete with a number of other heating systems. For this reason the cascade systems or even multistage compression used in refrigeration technology are not altogether acceptable in heat pump designs; hence the need to comply with the relevant demands made of the particular properties of the working fluid.

A working fluid should display a favourable thermodynamic equilibrium relationship between the pressure and temperature, and small pressure differences within the existing temperature range. One demands of the fluid a large enthalpy of evaporation and a moderate specific volume, so that the installation is not too large. The specific heat of fluid, both in its liquid and vapour phase, should be small to minimize losses resulting from the cooling of superheated vapour as well as from throttling the liquid. The boiling point temperature should be suitably low, to avoid negative gauge pressures in the installation. The specific enthalpy should be suitably high, so that a small mass flow of the working fluid will circulate in the plant. Trouton's rule results in the following approximate formula for the enthalpy of evaporation:

$$r = 88\ T_{bo}\ \text{kJ/kmol} \tag{2.1}$$

which means that fluids having small molecular masses are desirable. Apart from very few fluids, including water and ammonia, molecular masses of most of the relevant fluids are of the order of 100, and hence their specific enthalpies are numerically equal to the standard boiling point temperatures.

The properties of working fluids are collected in three tables. The first two tables present halocarbons derived from aliphatic hydrocarbons. These compounds are manufactured, and their properties are known, enabling the selection of materials which can be used in their presence (including the lubricating oils); thus a prolonged, trouble-free operation of plant can be guaranteed. Also significant is the fact that these compounds are well advertised, well marketed and sold through extensive dealership networks, and have established a good tradition in the market for many years. However, as wide-ranging changes have recently been announced in this group of working fluids, they are given somewhat more attention and are presented in two tables. The first (Table 2.1) presents compounds presently used, but not complying with current environmental requirements and thus subject to phasing out. The second (Table 2.2) presents halocarbons derived from aliphatic compounds which comply with the environmental protection requirements and are either already known and used, or being developed and prepared for marketing. The third (Table 2.3) presents other working fluids, which in some respects are superior to halocarbons derived from aliphatic compounds, although not as commonly used as the latter ones.

The halocarbon derivatives of aliphatic compounds are commonly used in heat pumps, but nearly exclusively in pumps of low power. Following the resolutions of the Montreal Convention of September 1987, one can expect changes in this group of working fluids. Hence some detailed information on these compounds is appropriate. The halocarbon derivatives are obtained by replacing hydrogen in the molecules of hydrocarbons – methane CH_4, ethane C_2H_6 and higher – with monovalent elements, chlorine Cl and fluorine F. These substitutions lead to such compounds as CH_2ClF, $CHCl_2F$, $CHClF_2$, $C_2HCl_3F_2$, $C_2Cl_3F_3$ etc. The compounds are allocated numbers according to a specific code. As the molecule can contain certain number of atoms of carbon (α), hydrogen (β), chlorine (γ) and fluorine (δ), their generalized chemical formula can be written as:

$$C_\alpha H_\beta Cl_\gamma F_\delta$$

The numerical symbols are obtained from the above formula, with consecutive digits calculated as follows:

$$(\alpha - 1)(\beta + 1)\delta$$

and if $(\alpha - 1) = 0$, the numerical symbol does not include a leading zero, thus the derivatives of methane have two-digit symbols, and derivatives of ethane have three-digit symbols. Thus the compound $CHCl_3$ has numerical symbol 20, $CHClF_2$ has 22, $C_2Cl_2F_4$ has 114, etc. Since all the fluids used in refrigeration are

Table 2.1 *Chlorofluorocarbons and azeotropic mixtures of CFCs commercially available and used but the use of which will be curtailed*

Numerical symbol	Kind of working fluid — Name, chemical symbol	Molecular weight	Standard boiling temp. (K)	Critical parameters T_{cr} (K)	Critical parameters P_{cr} (MPa)	Characteristic properties ODF*	Characteristic properties GWP**	Characteristic properties Toxicity	Characteristic properties Flammability
R12	Dichlorodifluoromethane CCl_2F_2	120.93	243.4	385.2	4.11	1.0	3.0	Low	None
R114	1,2-Dichlorotrafluoroethane $CClF_2CClF_2$	170.94	276.8	418.9	3.26	0.7	4.0	Low	None
R11	Trichlorofluoromethane CCl_3F	137.8	297.0	471.2	4.40	1.0	1.0	Low	None
R113	Trichlorotrifluoroethane CCl_2FCClF_2	187.39	320.8	487.3	3.37	0.9	1.4	Low	None
R502	48.8% R22, 51.2% R115	111.64	227.6	636.5	4.27	0.6	n.d.	Low	None
R500	73.8% R12, 26.2% R152a	99.31	239.7	378.7	4.42	0.7	n.d.	Low	None
R505	78% R12, 22% R31	103.43	243.6	391.0	4.73	0.8	n.d.	Low	None
R506	55.1 R31, 44.9% R114	93.69	260.8	415.4	5.16	0.4	n.d.	Low	None

* Ozone Depletion Factor
** Global Warming Potential

Table 2.2 *Chlorofluorocarbons: environmentally safe, available, under development, and proposed*

Kind of working fluid		Molecular weight	Standard boiling temp. (K)	Critical parameters		Characteristic properties				Remark
Numerical symbol	Name, chemical symbol			T_{cr} (K)	P_{cr} (MPa)	ODF	GWP	Toxicity	Flammability	
R22 (used temporarily)	Chlorodifluoromethane $CHClF_2$	86.48	232.4	369.2	4.97	0.05	0.30	Low	None	Available
R134a	1,1,1,2-Tetrafluoroethane CH_2FCF_3	102.00	246.7	374.3	4.06	0.0	0.26	Low	None	Under development to replace R12[21,22]
R152a	1,1-Difluoroethane	66.05	248.2	386.7	4.49	0.0	0.03	Low	Slight	Available
R134	1,1,2,2-Tetrafluoroethane CHF_2CHF_2	102.00	253.5	387.2	3.43	0.0	n.d.	Low	None	Proposed
R124 (used temporarily)	2-Chloro-1,1,1,2-tetra-fluoroethane $CHClFCF_3$	136.48	261.2	418.9	3.34	0.02	0.1	Low	None	Under development
R124a (used temporarily)	1-Chloro-1,1,2,2-tetra-fluoroethane CHF_2CClF_2	136.48	263.0	399.9	3.71	<0.1	n.d.	Low	None	Proposed

R142b (used temporarily)	Chlorodifluoroethane CH_3CClF_2	100.50	263.4	410.3	4.12	0.06	0.36	Low	Slight	Available
RC318	Octafluorocyclobutane C_4F_8	200.40	267.4	388.5	2.78	0.0	n.d.	Low	None	Available
R160	Ethyl chloride CH_3CH_2Cl	64.52	285.6	460.4	5.27	<0.1	n.d.	Mode Rate	Slight	Proposed
R123 (used temporarily)	Dichlorotrifluoroethane $CHCl_2CF_3$	152.91	300.8	458.2	3.73	0.02	0.02	Low	None	Under development to replace R11[21,22]
R141b (used temporarily)	1,1-Dichloro-1-fluoro-ethane CH_3CCl_2F	116.90	305.2	482.6	4.38	0.09	0.1	Low	Slight	Proposed
R280	Propyl chloride C_3H_7Cl	78.54	319.8	503.2	4.58	<0.1	n.d.	Low	None	Proposed
R150a	1,1-Dichloroethane	98.96	330.2	523.1	4.59	<0.1	n.d.	Low	Slight	Proposed

Table 2.3 List of other working agents in order of potential application

Numerical symbol	Kind of working fluid — Name, chemical symbol	Molecular weight	Standard boiling temp. T_{bo} (K)	Critical parameters T_{cr} (K)	Critical parameters P_{cr} (MPa)	Characteristic properties ODF	GWP	Toxicity	Flammability
R717	Ammonia NH_3	17.03	239.9	406.2	11.62	0.0	None	None	Moderate / Slight
R718	Water H_2O	18.02	373.2	647.6	22.86	0.0	None	None	None
R630	Methylamine CH_3NH_2	31.06	266.5	430.1	7.46	0.0	n.d.	n.d.	High
–	Methanol CH_3OH	32.04	338.0	513.2	7.85	0.0	n.d.	Moderate	High
–	Isopropylmercaptan $(CH_3)_2CHSH$	76.16	321.5	512.1	4.35	0.0	n.d.	–	–
R600	n-Butane C_4H_{10}	58.13	272.7	425.2	3.79	0.0	n.d.	None	High
–	n-Pentane C_5H_{12}	72.15	309.2	469.7	3.37	0.0	n.d.	None	High

additionally prefixed by the letter R, we have R22, R114, etc. Additional symbols are used; for example uppercase letter C placed before the numerical symbol denominates a cyclic molecule, and lowercase letters (a,b) placed after the numerical symbol denote consecutive isomers.

Figure 2.1a illustrates how such compounds based on ethane can be developed (there are no reserve compounds based on methane). Figures 2.1b and 2.1c, on the other hand, illustrate their general properties and suitability as working fluids, expressed by their boiling points. These fluids are mainly needed for refrigeration purposes; thus isomers 142b and 152a are desirable, as they have lower boiling points. For heat pump use isomers 142 or 152 would have been more suitable. Apart from the properties illustrated in Figures 2.1b and c, these compounds are characterized by two other important indicators, which are a measure of their environmental harmfulness. One is the ozone depletion factor (ODF), which quantifies the harmful effect of a fluid on the ozone (O_3) layer in the stratosphere. Species fully saturated with chlorine and fluorine, characterized by a long lifespan of the molecule, which do not contain any hydrogen atoms, such as R11 (CCl_3F), R12 (CCl_2F_2) and R114 ($C_2Cl_2F_4$), are regarded as particularly harmful, while compounds containing hydrogen, such as R22 ($CHClF_2$), are thought to be less harmful, and compounds which do not contain any chlorine, such as RC318 (C_4F_8), or the two compounds to replace R12. This harmful effect is determined according to a hypothesis of ozone depletion in the upper stratosphere which is not yet fully proven. Nevertheless, the Montreal Convention resolutions are mandatory and define the trends for developments in the halocarbon derivatives of aliphatic compounds. The second indicator is a measure of the degree of danger caused by the presence of a compound in the atmosphere, with regard to its contribution to the increase of the earth's surface temperature by impeding the radiation heat loss. This indicator is called the global warming potential (GWP) and is shown for various compounds in Figure 2.2.

Refrigerant R22 is characterized by a relatively low ratio of pressures in the high-temperature heat and low-temperature heat sources (low compression) and high enthalpy of vaporization per unit volume. It has, however, relatively high pressures, so it can be used without problems only in installations of small dimensions, and thus of low power. Since the number of such small installations is the largest, this working fluid takes first place in the table, which does not mean it is ideal. It has a quite low heat transferability, and thus the evaporator and condenser are relatively large. It also has a low critical temperature, which limits its use at higher temperatures of the high-temperature heat source. R22 does not co-operate well with oils. It is one of the more common working fluids of the halocarbon derivatives of the aliphatic compounds group. Its harmfulness to the ozone layer is regarded as low, so its production and use will not be limited.

The R12 refrigerant has properties similar to those of R22, but it has lower pressures, and thus it is being used in installations of larger dimensions and powers. It is the most common working fluid, but is regarded as harmful. It will be replaced by the new working fluids R134a or R152a.

The RC318 refrigerant is characterized by much lower pressures than R22 and R12, but it also has lower enthalpy of vaporization per unit volume. Its ability to transfer heat is also low. All in all, this working fluid is not too common, but as it

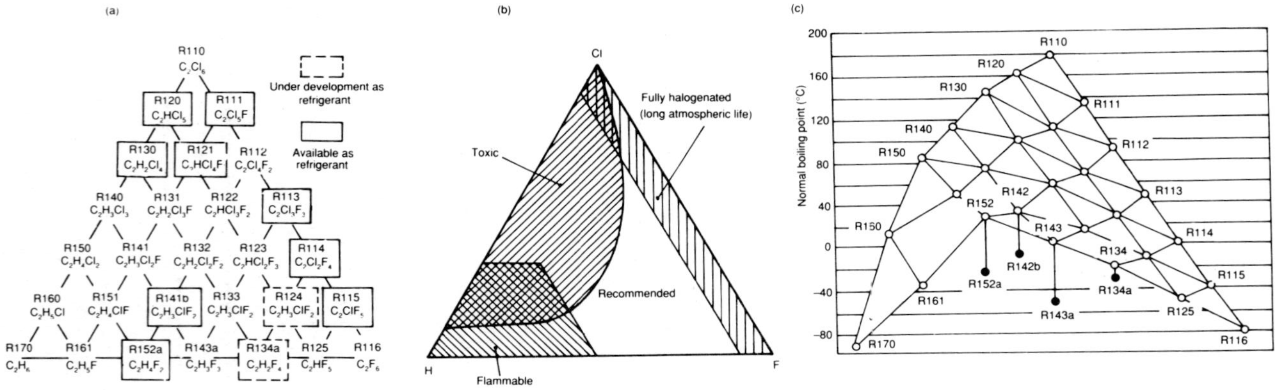

Figure 2.1 *General possibilities of obtaining halogenated chlorofluorocarbon derivatives of aliphatic compounds by substituting consecutive hydrogen atoms with chlorine or fluorine, and some of their properties: (a) possible working fluids based on ethane for use in refrigeration and heat pumps – those in solid box, available on the market; those in dashed box, presently under development; (b) general properties in relation to molecular structure; (c) standard boiling points; black points denote isomers [23]*

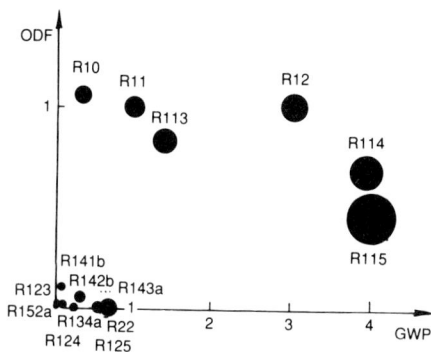

Figure 2.2   *Properties of halogenated chlorofluorocarbon derivatives of aliphatic compounds with respect to their harmful effects on atmosphere and illustration of new advantageous properties sought [21].*

does not contain chlorine, it is completely harmless towards the ozone layer. For this reason one can expect an increase in its popularity.

The R114 refrigerant has thermodynamic properties similar to those of RC318, but much better heat transfer abilities. It is, however, regarded as harmful.

The R21 refrigerant has relatively low pressures, but a high ratio of high to low pressure and low enthalpy of vaporization, so that it requires the use of turbocompressors. It is corrosive and harmful to the ozone layer.

The R11 refrigerant has properties similar to those of R21; it has very good heat transfer parameters. It is corrosive and regarded as harmful; it is to be replaced by the R123.

The R113 refrigerant has low pressure, so it is used in turbocompressors. It has poor heat transfer parameters and is regarded as harmful.

The R502 refrigerant is an azeotrope containing 48.8 mass per cent R22 and 51.2 mass per cent R115; its properties are similar to those of R22.

Zeotropic blends of working fluids, e.g. 0.55 R32 + 0.45 R15a, or 0.67 R32 + 0.33 R134, are also used to provide condensation and evaporation processes in non-isothermal conditions. Such blends enable better proximity of the condensation and evaporation temperatures to the temperature of substances in the sources, thus limiting exergy losses in the exchangers.

The R123, R134a and R152a working fluids are not as yet manufactured on an industrial scale. These do not contain chlorine and are harmless to the ozone layer. They are meant to replace: R123, R11 and R134a or R152a, and R12. The R152a is flammable, so that R134a has better prospects.

Ammonia (NH_3) has been used in storage refrigeration for years; it is, however, toxic and in certain concentrations even explosive, and also it has a characteristic unpleasant odour. These properties eliminate it from use in domestic heat pump installations; nevertheless it is used in large-power compression and sorption installations with water as a sorbent. The relatively high pressure of ammonia at higher temperatures limits its use.

Water (H_2O) is a fluid displaying many virtues: high enthalpy of evaporation, good heat transfer properties, availability, cheapness, etc. It also has some drawbacks: it cannot be used in temperatures below 0°C; and at low temperatures it has a very high specific volume, which hinders its use in compression heat pumps, particularly with piston compressors. However, it is used in sorption heat pumps. In compression cycle heat pumps water is used when low-temperature heat source temperatures are above 320 K, this means various kinds of plant utilizing waste heat, of large scale and with compressors other than reciprocating piston designs (e.g. screw compressors).

Methylamine (CH_3NH_2) is used only sporadically in research installations. It has low pressure at the low-temperature heat source, but high pressure at the high-temperature heat source. It is used in sorption cycle heat pumps with water as the solvent. It is toxic and has a characteristic unpleasant odour.

Methanol (CH_3OH) is also used sporadically, only in sorption cycle heat pumps, with water as the solvent; its characteristic is a low pressure at the high-temperature heat source. It is toxic.

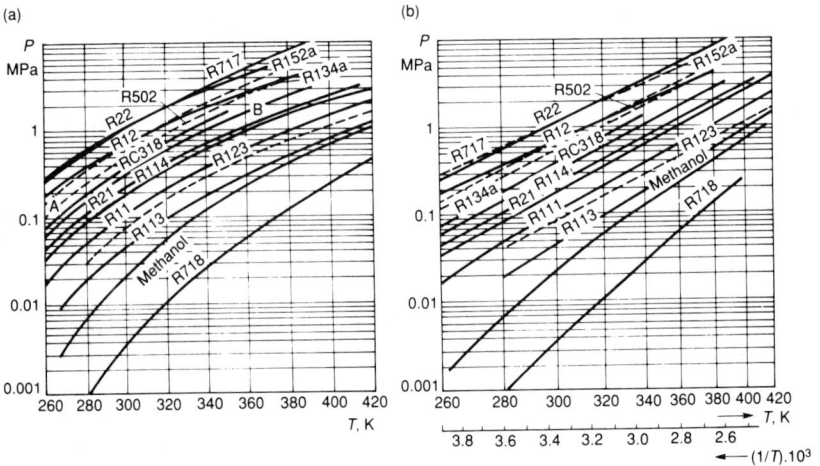

Figure 2.3 *Pressure as a function of temperature (at saturation):* (a) *in log P–T coordinates;* (b) *in log P–1/T coordinates*

Figure 2.3 shows the saturation pressure lines of $P = P(T)$ for the more important working fluids (in the log P–$1/T$ coordinates this relationship is nearly linear). High pressures are undesirable, as this increases the cost of plant, and so are negative gauge pressures, as there might be problems with leaks. The equilibrium line is not sufficient to allow the choice of a working fluid, but it can be used to present a logical procedure suggesting such a choice. Let us assume that we are choosing a working fluid for a compression cycle heat pump with a piston compressor having a compression ratio of 10. The temperature of the low-temperature heat source is 280 K; the pressure range in the installation is 0.1–1 MPa. For these conditions, the highest temperature that can be achieved in the high-temperature heat source, 360 K, can be achieved for the working fluid R114. This working fluid, however, has quite large specific volume, and quite low values of the heat of evaporation. This is illustrated (after [24]) in Figure 2.4, from which one can conclude that for the three working fluids shown, R22, R12 and R114, a compressor using R22 would have the smallest dimensions, and the highest temperature in the high-temperature heat source can be achieved for R114.

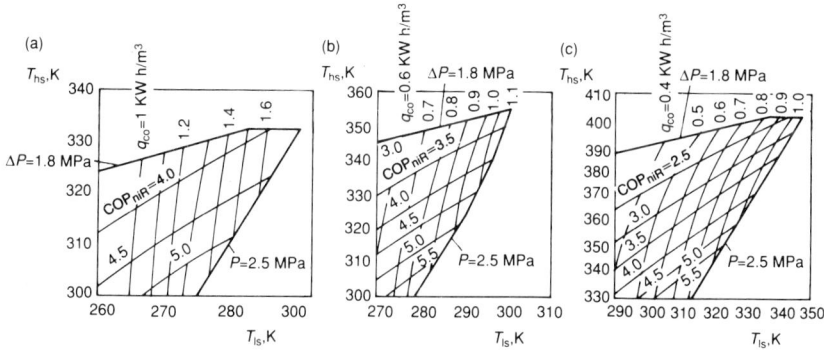

Figure 2.4   *Temperature range in a compression cycle heat pump, obtained with the use of different working fluids:* (a) *R22;* (b) *R12;* (c) *R114*

Two other aspects of the location of the $P = P(T)$ line are worth further consideration. First of all, all these lines have approximately the same slopes. This means that there are no sensational working fluids having a slope of the equilibrium line allowing us to reach point B when starting from point A (Figure 2.3). Secondly, attractive temperatures for heat pumps are in the region of 400 K and above. Thus methanol and water seem to be more promising for these temperatures than typical working fluids used in refrigeration. One cannot, however, obtain large compression ratios for methanol and water (in single-stage piston compressors) and thus the high-temperature heat source temperatures for compressor based heat pumps are limited.

Figure 2.5 illustrates the changes of specific volume of the working fluids. The specific volume is an important parameter in a compression cycle heat pump, as it does to some degree determine the volume of the compressor. Large specific volumes of the fluid are unacceptable in piston compressors, but can be dealt with in centrifugal compressors, absorption cycle heat pumps or heat transformers. One can deduce from Figure 2.5 that the volume of necessary compressor decreases with the increase of the condensation temperature, which is very attractive for water as a working fluid. It is worth mentioning here that water as a working fluid is gladly used in absorption cycle heat transformers, where the cost of mechanical pumping of the fluid is relatively small. Another particular characteristic of a working fluid is the path of the equilibrium line in the T–s plane (Figure 2.6). This path enables us to determine the distance of processes in the cycle from the critical point, and also the heat of evaporation. The temperature of the cycle not only cannot exceed the

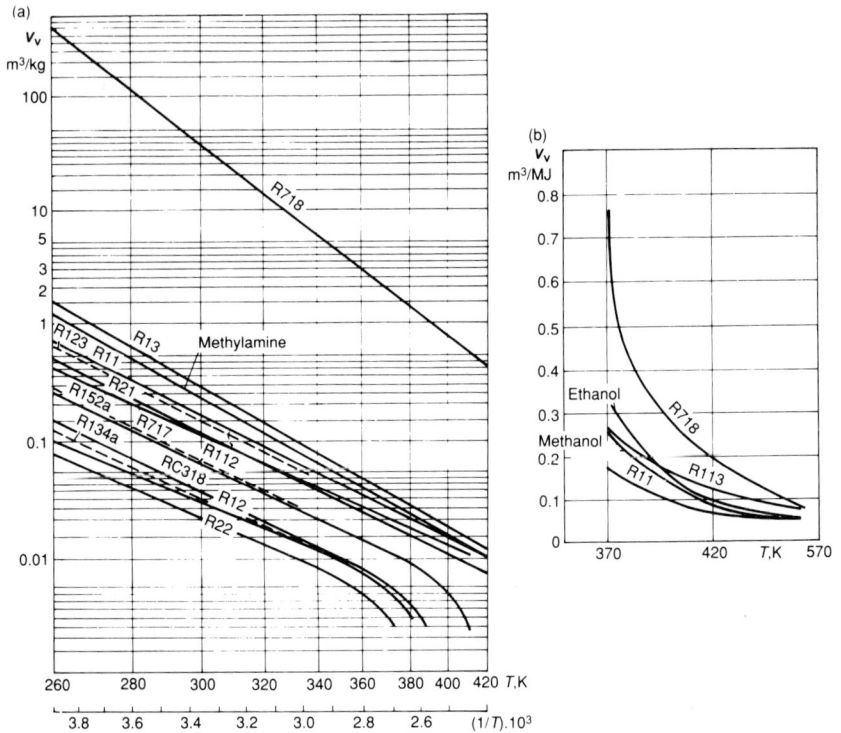

Figure 2.5 *Specific volume versus temperature (at the saturation line): (a) in v–T coordinates; (b) ratio of the change of fluid volume to the enthalpy of evaporation as a function of temperature*

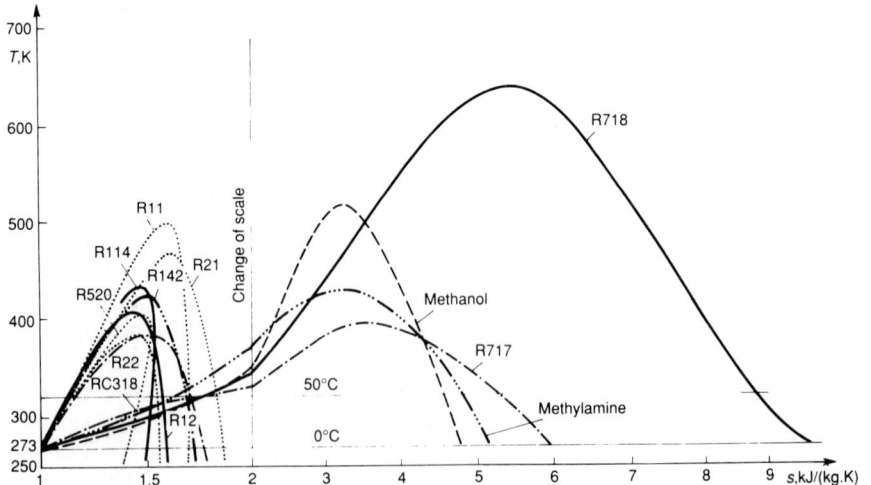

Figure 2.6 *Curves representing the boiling liquid and saturated vapour in T–s coordinates*

critical temperature, but should not approach the latter too closely, as the cycle then becomes less effective (Figure 2.7b). It is recommended that $T_{ev}/T_{cr} < 0.75$. The heat of evaporation is proportional to the surface area limited by the relevant isotherm (isobar) as is shown in Figure 2.7a. Generally speaking, this heat should be large, so that little fluid needs to circulate in the installation. An important parameter here is the ratio of the heat of evaporation to the volume of fluid; this ratio should be as large as possible.

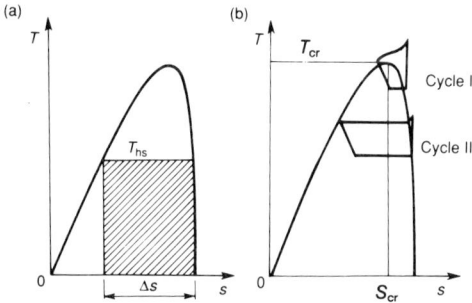

Figure 2.7   *Curves representing boiling liquid and saturated vapour in T–s coordinates:* (a) *interpretation of the heat of condensation T_{hs}, ΔS;* (b) *change of cycle performance; I, close to the critical point; II, sufficiently far from the critical point*

The properties of fluid relating to its ability to transfer heat can be shown for conditions prevailing either in the condenser or in the evaporator. Assuming that the condenser consists of water-cooled horizontal non-finned tubes, one can determine the role of physical properties of the fluid in the heat transfer coefficient (on the vapour side). Employing to this effect the Nusselt theory, which yields good results for the condensation of various vapours, one can identify a measure a of this role in the form of a relationship $a = k^{3/4}\rho_L^{1/2}r^{1/4}/\mu^{1/4}$. This quantity is different for various working fluids and changes with temperature (Figure 2.8a). Taking into account the fact that the kinetics of condensation processes depend mainly on the value of a, then, for example, if the ratio of these values for water and the R12 working fluid is seven, we can (without entering into any detailed analysis of heat transfer) say that a condenser for water will be approximately 2.5 times smaller than a condenser for R12. These facts are the reason behind the research into the intensification of condensation processes of the halocarbon compounds. By properly shaping the condensation surface one can introduce the surface tension forces, which enhance the heat transfer process, and in effect the above-mentioned differences in the condenser sizes can be eliminated. These forces can be introduced by using tubes finned in a specific way, enabling the introduction of surface tension forces to the condensate flow. Recently attempts have been made to introduce electrical fields into the condenser volume, which also leads to the intensification of heat transfer.

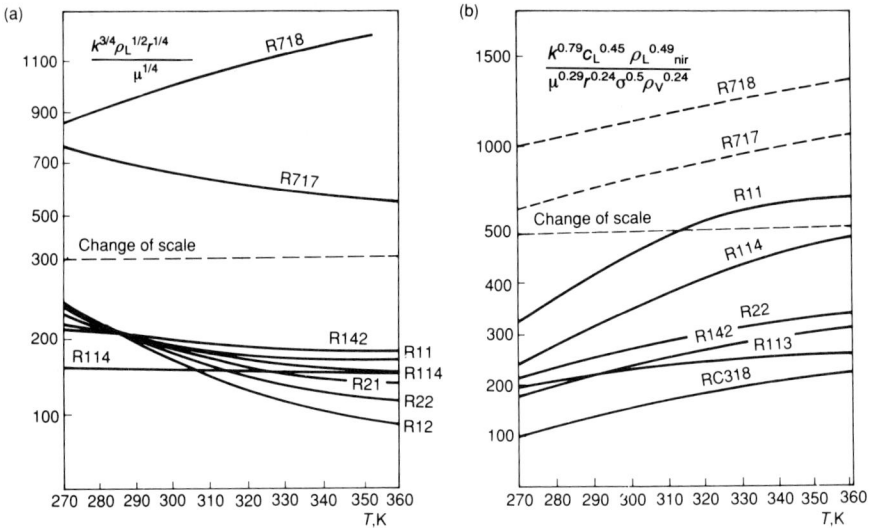

Figure 2.8 *Comparison of fluid properties at:* (a) *condensation in a condenser;* (b) *boiling in an evaporator. Both are of a shell and tube design with horizontal tubes*

In present conditions the R114 working fluid is a very good working fluid for the compression cycle heat pumps characterized by high-temperature heat source temperatures in the region of 400 K, and water is very good for the sorption cycle pumps. There is a lack of good fluids for the compression cycle heat pumps in the temperature interval between 400 and 470 K. At temperatures higher than 470 K, water becomes a very good working fluid in compression cycle heat pumps. These pumps are characterized by high COPs and large temperature differences between the high- and low-temperature heat sources. For example, at a temperature difference of 70 K and high-temperature heat source temperature of 470 K, the COP is around 5. The result of investigations of a heat pump using water as a working fluid are reported in [25].

2.2 Sorbents and their specification

Sorbent is a liquid used for absorption of the working fluid (we have omitted here the adsorption-based cycles, where sorbent is used as a solid body, because of their marginal applications). The sorbent, and perhaps also the salts used for their preparation, should have certain general and particular characteristic properties. The demands on general properties are the same as for the working fluids, and particular characteristics relate to the suitability of a sorbent to a practical realization of a cycle. These particular demands are, essentially, provision of good absorption of the working fluid and easy separation of the vapours of the working fluid (desorption) from the sorbent. The sorbent has a higher boiling point temperature (at a given pressure) than the working fluid. The differences in temperatures and partial pressures decrease as the solution is being diluted, but in such a way that the appropriate mutual concentration of sorbent and working fluid in the solution can be maintained. The properties of a sorbent formed from a solution of the liquid phase of working fluid and the salt are similar to the properties of a solution being a mixture of the working fluid and the liquid sorbent. The advantage of sorbent formed from salt solutions is a very small partial pressure of salt above the surface of solution, allowing the desorption of working fluid from this solution to be achieved through distillation alone (there is no need to use rectification).

The thermodynamic equilibrium of a solution, i.e. a relationship between the concentration, pressure and temperature, can be presented in various systems of coordinates, illustrated in Figure 2.9. The systems are different from each other, but each of these systems highlights a particular characteristic of the equilibrium. Therefore the $T = T(z)$ plot illustrates (Figure 2.9a) the range of boiling point temperatures dependent on concentrations of the working fluid in the solution (x)

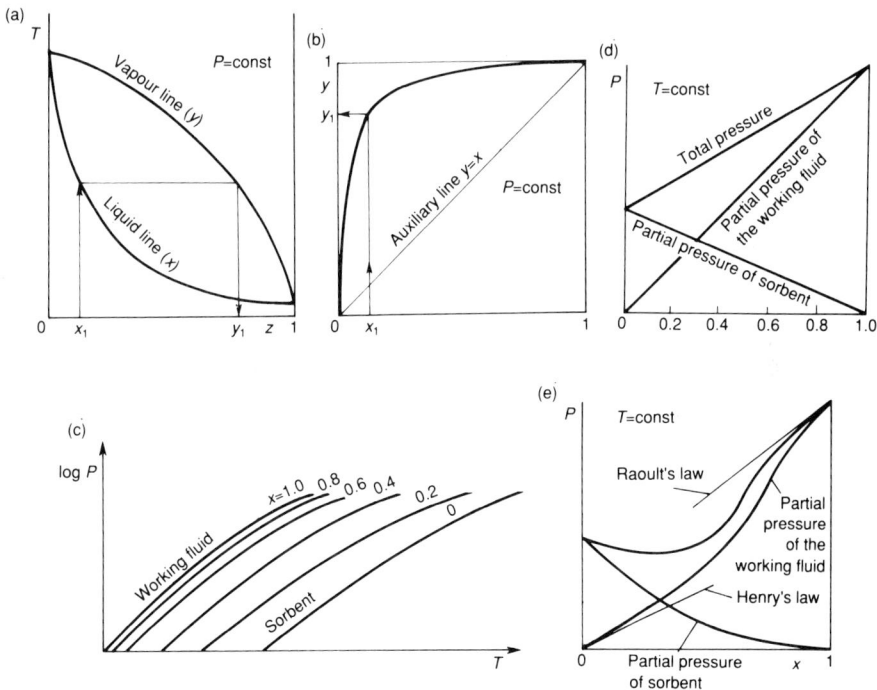

Figure 2.9   *Thermodynamic equilibrium of a two-component solution (limited to fully miscible liquids): (a) temperature as a function of concentrations; (b) concentration in the vapour as a function of concentration in the liquid; (c) pressure as a function of temperature; (d) total vapour pressure (the sum of partial pressures) as a function of concentration for an ideal solution; (e) pressures for a non-ideal solution with a negative deviation from Raoult's law*

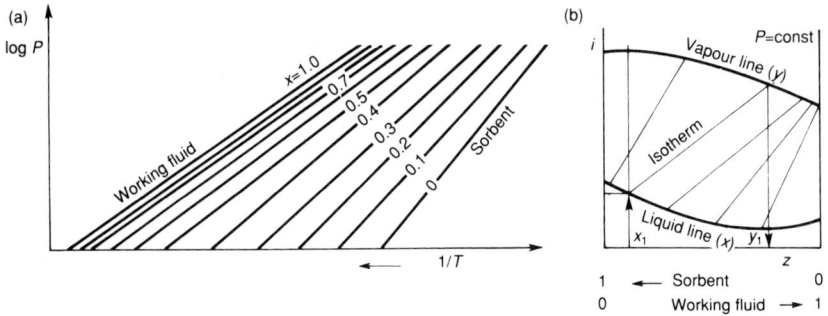

Figure 2.10 *Diagrams used in quantitative analysis of thermodynamic cycles of the sorption cycle heat pumps showing isotherms in the wet vapour region:* (a) *log* $P = f(1/T)$, $z = const.$; (b) $i = i(z)$

and vapour (y). The $x = x(y)$ plot, shown in Figure 2.9b, helps us to interpret the separation of vapours of the working fluid from those of sorbent. Also, the plot of $P = P(T)$ at $x = const.$ (Figure 2.9c) can be used for this interpretation. The last plot is a generalization of the plot shown in Figure 2.1 for the case of a solution of two miscible liquids or a solution of salt in a liquid. Figure 2.9d shows partial pressures of vapours of the sorbent and working fluid and total pressures of an ideal solution. Figure 2.9e shows the negative deviation from Raoult's law. A negative deviation from Raoult's law is demanded from a solution used in a heat pump, as this leads to smaller flow rates of solution circulating between the absorber and desorber.

For quantitative analysis one finds that the plots of $P = P(1/T)$, $x = const.$ and $i = i(z)$, shown in Figure 2.10, are more convenient. First of these enables determination of the intensive parameters of the cycle: P, T, z. This plot is similar

Table 2.4 *Values of the directional coefficient for the slope of the line* $P_s = T(T_s)$ *in the log* P–$1/T$ *coordinates*

Working fluid	Values of the directional coefficient $m = -rT/(P \, \Delta v) \times 10^3$ K	
	At a pressure corresponding to temperature 273 K	Close to the critical point
R22	1.0609	1.0297
R12	1.0650	1.0140
RC318	1.2510	1.2460
R11	1.3910	–
R717	1.2120	1.1770
R718	2.3430[a]	2.2660

[a] At a pressure corresponding to temperature $T_s = 279.948$ K.

to that of Figure 2.9c; it differs in the temperature scale. With such temperature coordinates the lines $z = $ const. in Figure 2.10a, both for the working fluid and for the solution, are straight for the whole range of pressures and temperatures, up to the critical point. Strictly speaking, this is not true, as the derivative $d(\log P)/d(1/T) = - Tr/(P\Delta v)$ does not have a constant value for each point, but changes slightly (Table 2.4). The plot $i = i(z)$ is particularly useful for interpretation of the heat effects related to the heat pump cycle. It is used in all sorts of calculations of the heat pumps, and previously it was used for sorption cycle refrigeration equipment, where it was introduced by Merkl and Bosniakovic in the 1930s. In principle it is valid only for constant pressure, as is shown in the i–x plane (with pressure as a parameter). The plot enables graphical calculations of the cycle of a sorption heat pump having two pressure levels, although some of the processes are projected onto the plot, which makes their physical interpretation difficult. The sorbents and salts used for their derivation are listed in Table 2.5. There are not many of these, but surely the list is not complete. Some of the sorbents have been in use for a long time, e.g. water solution of lithium bromide. Some others are only used in a few research installations. Among these there are substances, such as water, used as a working fluid or used to collaborate with one or many fluids.

Table 2.5 *Physicochemical properties of exemplary sorbents used in the sorption cycle heat pumps*

Sorbent	Lithium bromide	Dimethyltetraethylene glycol[a]	Sodium rhodate	Lithium nitrate
Chemical formula	LiBr	$CH_3OCH_2(CH_2)_4OCH_3$	NaSCN	$LiNO_3$
Molar mass	86.85	222.30	81.07	68.95
Solidification (melting) point[b] (K)	–	245.0	–	–
Boiling point[b] (K)	1538	548.45	–	Dissociates at 873 K before boiling
Polarity	Very poor	Poor	Poor	Poor
Chemical stability	Good	Good	Good	Good
Corrosiveness	High	Very low	–	–
Toxicity	Very low	Very low	High	–

[a]Trade name E181.
[b]At atmospheric pressure.

The working fluids and sorbents used in heat pumps are listed in Table 2.6, roughly in order corresponding to their range of applicability. Figure 2.11 shows diagrammatically the pressure and temperature ranges possible in a heat pump cycle when using a particular fluid–sorbent pair. Large temperature differences, relatively low pressures and gentle slopes of lines are advantageous in order to

Table 2.6 *Examples of matched pairs of working fluids and sorbents used in sorption cycle heat pumps*

Working fluid	Sorbent (solvent)	Solubility limit z_{lim}
Ammonia	Water	At all concentrations
Water	Lithium bromide	0.35
R22	E181[a]	At all concentrations
Methanol	Water	As above
Methylamine	Water	As above
Ammonia	Lithium nitrate	0.30
Ammonia	Sodium rhodate	0.35

[a]Trade name – dimethyltetraethyleneglycol.

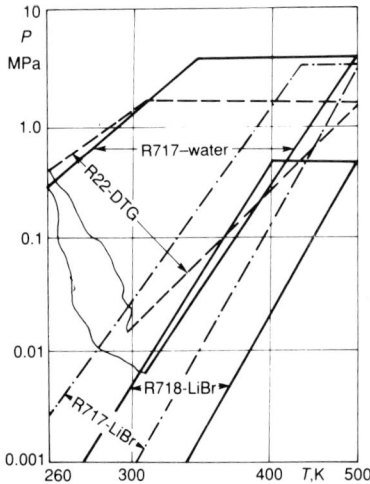

Figure 2.11 *Ranges of temperatures and pressures used for different pairs*

avoid high compressions, although high compression ratios are not as important here as they are in compression cycle heat pumps. The matched pair R717–water displays quite good properties and thus it has been used for years, particularly in refrigeration. The equilibrium data for this pair are shown in Figure 2.12. Relatively high pressures in this case are a serious disadvantage. Somewhat better properties are displayed by the pair R22–DTG (dimethyl ether of tetraethylene glycol). Relatively good properties, particularly in the higher temperature range, are those of the pair R718–water solution of lithium bromide; the equilibrium data are shown in Figure 2.13. This pair is used at a rather higher temperature range $T_{hs} = 390\,K$, and sometimes even up to $T_{hs} = 430\,K$. Two weaknesses of this pair are: very low pressures of the working fluid at temperatures close to the

Table 2.7 *Examples of salts of possible use in sorption cycle heat pumps*

Salt		Property			
Metal	Halogen	Molar heat of dissolution Δi_d (kJ/mol)	Solubility[a] (%)	Melting point T_m (K)	Negative deviation from Raoult's law ΔP_R (kPa)
Li	I	61.70	61.200	719	3.79
	Br	47.50	61.000	820	3.47
	Cl	35.00	45.350	887	3.42
	F	−4.34	0.270	1143	–
Ca	I_2	117.54	67.600	848	–
	Br_2	102.40	58.800	1033	5.53
	Cl_2	72.77	42.700	1045	5.07
	F_2	11.28	0.016	1603	–
K	I	21.35	59.000	996	–
	Br	−21.23	39.500	1003	3.19
	Cl	−17.50	25.400	1049	3.20
	F	15.00	48.700	1153	–
Na	I	5.10	64.100	924	3.43
	Br	−0.79	47.600	1028	3.39
	Cl	−4.90	26.370	1073	3.36
	F	−2.50	3.900	1265	–
Mg	I_2	208.10	58.300	Sublimes	–
	Br_2	180.99	50.270	973	5.67
	Cl_2	150.14	35.300	985	5.17
	F_2	11.62	–	1669	–

[a]At temperature T_{en} = 291 K.

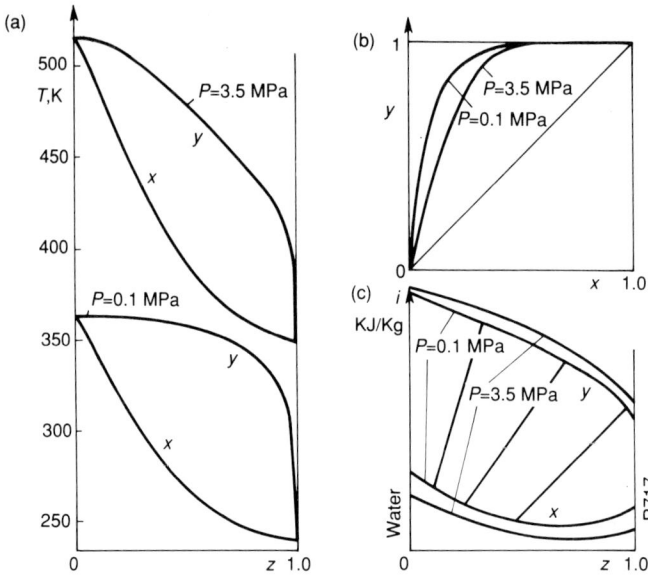

Figure 2.12 *Equilibrium data for the pair R717–water:* (a) $T = T(z)$; (b) $y = y(x)$ *(of little use for x > 0.5);* (c) $i = i(z)$

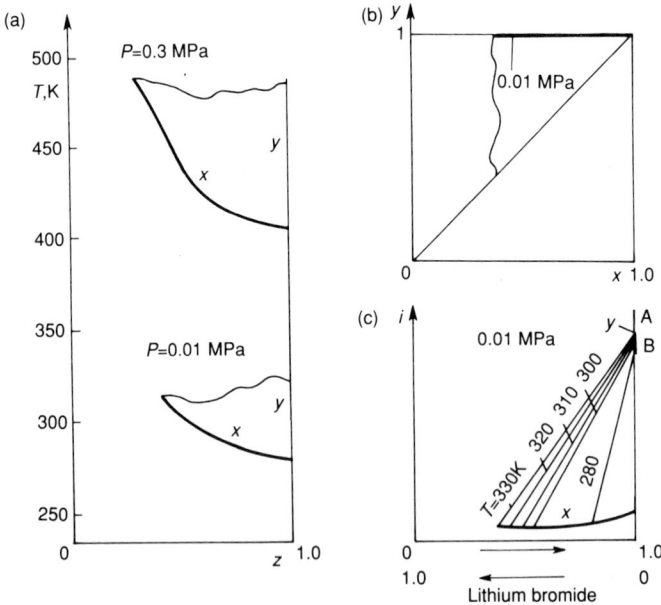

Figure 2.13 *Equilibrium data for the pair R718–water solution of lithium bromide;* (a) $T = T(z)$; (b) $y = y(x)$ *(of little use because the slope of the y = y(x) line is difficult to illustrate);* (c) $i = i(z)$ *with the vapour line degenerated over the distance AB*

environmental range, and corrosiveness of the sorbent at higher temperatures, which is the reason behind research into other solutions of metallic salts [27]. Table 2.7 lists the recently proposed [28] types of these salts and some of their properties. The analyses of various combinations of working fluids and sorbents are given for example in [28], [29], [30] and [31].

The plots of log P–$1/T$ and $i = i(z)$ for some of the most commonly used combinations are given in the Appendix, together with the equations describing their properties in algebraic form.

Chapter 3

The high- and low-temperature heat sources

3.1 General information on sources

A heat pump cycle (Figure 1.1) includes two sources: the low-temperature heat source from which thermal energy is absorbed and the high-temperature heat source to which it is discharged. In this way the high-temperature heat source somewhat determines the useful effect of a heat pump, and the low-temperature heat source supplies the raw material in form of energy at a low exergy level, to be processed into energy at a higher exergy level. So far the description has been based on a simplified treatment of the sources. It has been assumed that the sources are isothermal and that any heat exchange taking place within the sources occurs in isothermal conditions. Issues relating to the absorption of energy at the low-temperature heat source level and its discharge for use at the high-temperature heat source level have been omitted.

Finite temperature differences within the sources and their effect on the heat pump cycles are explained qualitatively in Section 3.2 (quantitative treatment, relating to the determination of exergy losses is given in Sections 4.3 and 5.3). Additional information is given in [32] and other works.

The problems relating to high-temperature heat sources (discharge of energy) are quite standard, to such an extent that they do not require special discussion. Many kinds of low-temperature heat sources (where energy is absorbed) are possible and each of these warrants individual treatment.

The kind of low-temperature heat source is very important, not only for the definition of a particular heat pump, but also for a decision on its application. Discussion of the low-temperature heat source requires taking into account some of its characteristics describing the source in a qualitative sense, and also its quantitative parameters.

The qualitative characteristics of a low-temperature heat source are:

1. Availability
2. Corrosiveness

The majority of heat pumps presently installed worldwide are used for heating scattered individual residential houses. In such conditions only natural sources are available: atmospheric air, or less frequently soil or solar energy. Surface waters from rivers or lakes (even more so from the sea) are not accessible for small heat pumps, easier to access is underground water from a sunk or deep well. It is important that a heat pump installed to heat an individual house does not disturb, by drawing energy from the low-temperature heat source, the operation of any other heat pumps installed in the neighbourhood. In this situation one prefers to use sources which are independent from each other, such as atmospheric air or solar energy. Temperature levels of the high-temperature heat source eliminate, for economic reasons, the transmission of energy over any distance, particularly for small heat pumps. Therefore, the low-temperature heat source must be found at the location of the heat pump. Heat pumps which utilize artificial low-temperature heat sources are also located at the place of heat demand, and at the same time at the low-temperature heat source location. Only when the heat pump is a sufficiently large unit can it constitute an element of a larger energy system, e.g. co-operate with the hot water mains. The low-temperature heat source does not then have to be located at the place of energy discharge.

Some of the substances in low-temperature heat sources necessitate the use of corrosion-resistant materials in the construction of a heat pump. Among such substances are brackish waters from deep wells, sea water, and some waters from artificial low-temperature heat sources. Corrosion can also be caused by vapours from dryers, because of water condensation, particularly in the presence of such gases as SO_2 and CO_2. Because of the economic restrictions on one hand, and higher costs of corrosion-resistant materials on the other, it can become an important problem.

Parameters defining the low-temperature heat source in a quantitative sense:

1. Temperature and its variation with time
2. Energy reserves and their variation with time
3. Investment and running costs

The temperature of the source and its time variations are important characteristics affecting the COP of a heat pump. The higher the low-temperature heat source temperature, the higher the COP of the heat pump and the higher the usefulness of heat obtained at the high-temperature heat source level. The temperatures of natural sources depend both on the kind of source and on the season. Every artificial source, on the other hand, is characterized by a temperature resulting from some technological process producing the waste heat. Thus it is difficult to generalize. Each case is quite different and must be discussed separately. Generally, however, the level of these temperatures is quite high for a low-temperature heat source: 290–350 K, sometimes even higher. The temperatures of the artificial sources do not depend on the season; natural sources, on the other hand, show large variations of temperature. In order to somewhat generalize the issue, a concept of coherency or consistency of the source yield with the energy demand was introduced. Thus if the heat pump is used for heating dwellings, within the heating season the heat demand varies; it is smaller at the beginning and end of the season and largest in its middle. Alas, the majority of natural heat sources are non-coherent, i.e. their yield is lowest at the time of peak energy demand. Particularly non-coherent is the atmospheric air. Most coherent sources are provided by deep underground waters, and some by soil and river water. Their temperatures do not change as much as that of atmospheric air.

The costs of the low-temperature heat source consist of capital investment costs of its intake, and running costs, mainly resulting from the cost of power to drive pumps and fans. For natural sources the costs of capturing the atmospheric air are low (£800 for a 20-kW pump), the medium costs are those of capturing water from a well (twice as much) and from the soil (three times as much), and capturing the solar energy is expensive (five times as much). Very high costs are associated with deep drilling to capture thermal waters (up to £3000 per 1-m depth). It is difficult, on the other hand, to assess the costs of artificial sources, as each of these is different, but generally these costs are not insignificant.

3.2 Finite temperature differences in the low- and high-temperature heat sources, and effect of these differences on the cycle and its performance

Energy is delivered from the source or discharged into the source through a heat exchanger. Temperature differences between the substance of the source and the working fluid occur in such a heat exchanger as a result of the heat transfer process; also, a decrease of temperature of the source substance can occur because of its limited heat capacity. Temperature drops of the working fluid can also occur during its condensation and evaporation (in a vapour compression cycle employing a two-component working fluid and in resorption cycle heat pumps). The effect of finite temperature differences can be shown using an example of a vapour compression cycle (with a single component working fluid) using the theoretical non-ideal reverse Rankine cycle. Figure 3.1 shows the temperature differences in the heat exchangers of the high-temperature heat source (condenser) and low-temperature heat source (evaporator) as a function of the exchanger length (Figure 3.1a) and the corresponding non-ideal reverse Rankine cycle (Figure 3.1b);

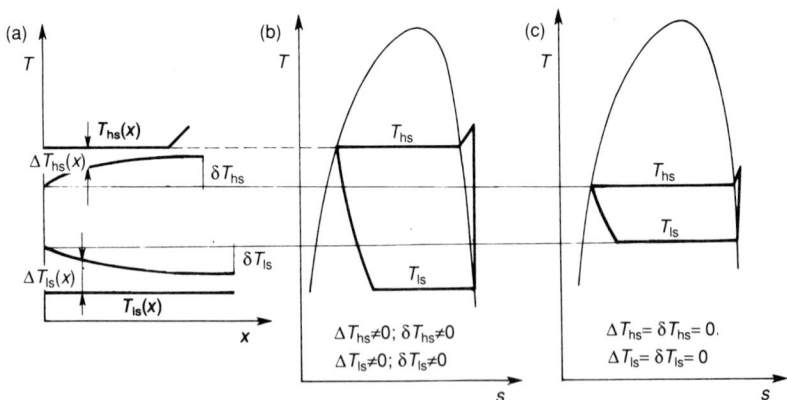

Figure 3.1 *Effect of finite temperature differences in the heat exchangers of high- and low-temperature heat sources in a vapour compression cycle heat pump:* (a) *temperature profiles along the length of heat exchangers in the high- and low-temperature heat sources;* (b) *the non-ideal reverse Rankine cycle for temperature differences as* (a); (c) *the non-ideal reverse Rankine cycle for idealized (isothermal) heat transfer conditions*

also shown is the non-ideal reverse Rankine cycle for idealized, isothermal heat transfer conditions with no temperature drops (Figure 3.1c).

Two characteristic temperature differences can be identified in each exchanger. One of these represents the driving force of heat exchange at each point along the exchanger ($\Delta T_{hs}(x)$; $\Delta T_{ls}(x)$). The average value of this driving force can be to some extent minimized by enlarging the heat transfer area. To ensure a suitable COP of a heat pump, this temperature difference must be kept at a relatively low level, not larger than a few degrees. The second one represents the changes of the actual temperature of the source substance in the exchanger (δT_{hs}, δT_{ls}); it is related to the finite heat capacity of the source substance flow through the exchanger. This temperature drop should also be minimized, and the only way to do it is to increase the source substance flow rate. If the source itself contains very large reserves of energy (e.g. a low-temperature heat source of atmospheric air or water from a river), then an increased flow rate of the source substance can be achieved by installing a suitably large pump or fan. This leads, however, to an increase in the capital and running costs, which reduces the economic effect of employing a heat pump. A large fraction of low-temperature heat sources, particularly of an artificial nature, contain only limited reserves of energy, a fact which to a large extent determines the temperature decrease of the source substance (δT_{ls}). An idea of a thermodynamic cycle, taking into account temperature changes of the source substances, is shown in Figure 3.2. A single Carnot cycle (Figure 3.2a) is replaced consecutively by three cycles (Figure 3.2b), 15 cycles (Figure 3.2c) and an infinite number of cycles (Figure 3.2d). With a larger number of Carnot cycles the heat pump will have a higher COP. The most popular vapour compression cycle of a heat pump employing a single-component working fluid is, then, not particularly advantageous when the source substances show large temperature changes. A much better choice in these conditions is a vapour compression heat pump employing a two-component fluid (Figure 1.5), where the composition of a working fluid is properly matched to the particular temperature drops. Sorption cycles also have advantageous properties in this respect, as the

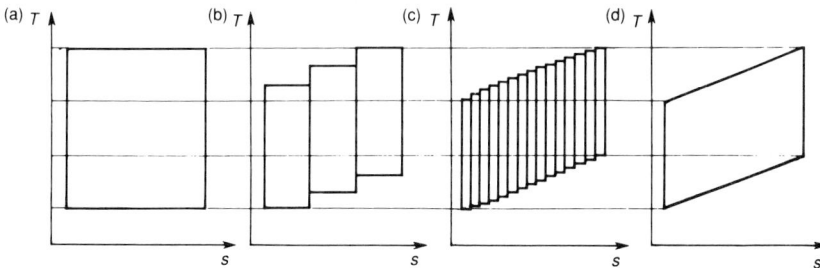

Figure 3.2 *Effect of finite temperature drop of the substance in the high-temperature (δT_{hs}) and low-temperature (δT_{ls}) heat sources on the coefficient of performance of a heat pump described using the ideal (Carnot) cycles: (a) single cycle ($COP_C = 4.9$); (b) three cycles ($COP_C = 6.5$); (c) fifteen cycles ($COP_C = 7.0$); (d) infinite number of cycles ($COP_C = 7.1$). For $\delta T_{ls} = \delta T_{hs} = 0$, $COP_C = 7.4$*

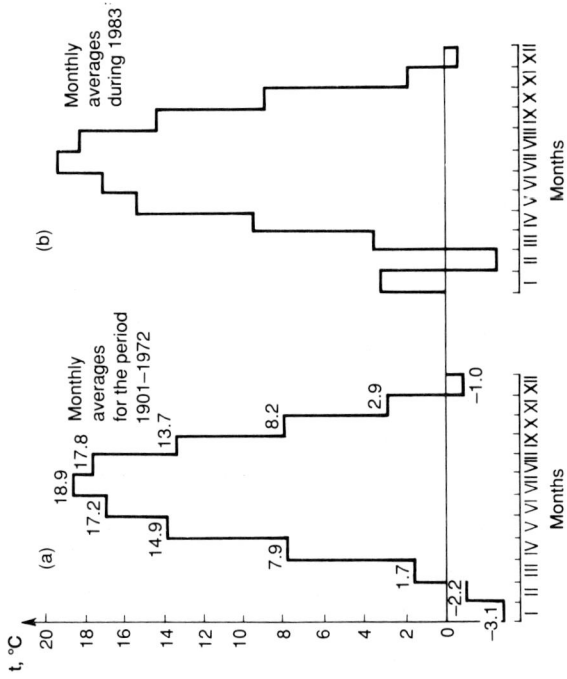

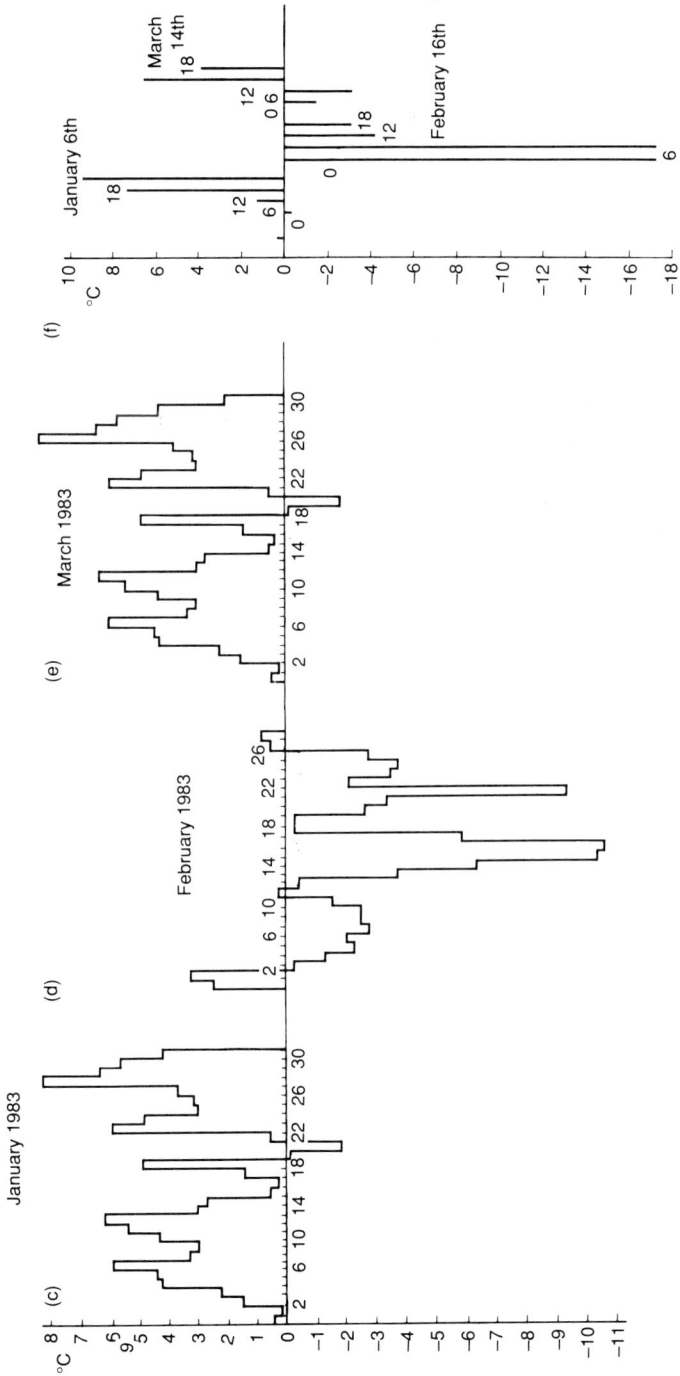

Figure 3.3 *Air temperature charts for Warsaw. monthly averages: (a) for the period 1901–1972; (b) during 1983. Daily averages: (c) January 1983; (d) February 1983; (e) March 1983; (f) variations during the day for selected days from (c, d, e). Temperatures measured at 0, 6, 12 h GMT*

absorption and desorption processes occur in non-isothermal conditions, which may be utilized to minimize the temperature differences between the working fluid and substance in the heat source.

3.3 The natural low-temperature heat sources

3.3.1 Atmospheric air

Atmospheric air is the most common supply source for small heat pumps. This fact results from its advantages: availability and low capital cost of building an installation for its intake. Figure 3.3 shows a yearly temperature profile for the city of Warsaw, based on the Monthly Hydro-Meteorological Bulletin [33].

One of the disadvantages of atmospheric air as a low-temperature heat source is its poor coherence (Figure 3.3) in a heat pump used for domestic heating. At relatively low-temperatures (close to °C), when the demand for heat at the

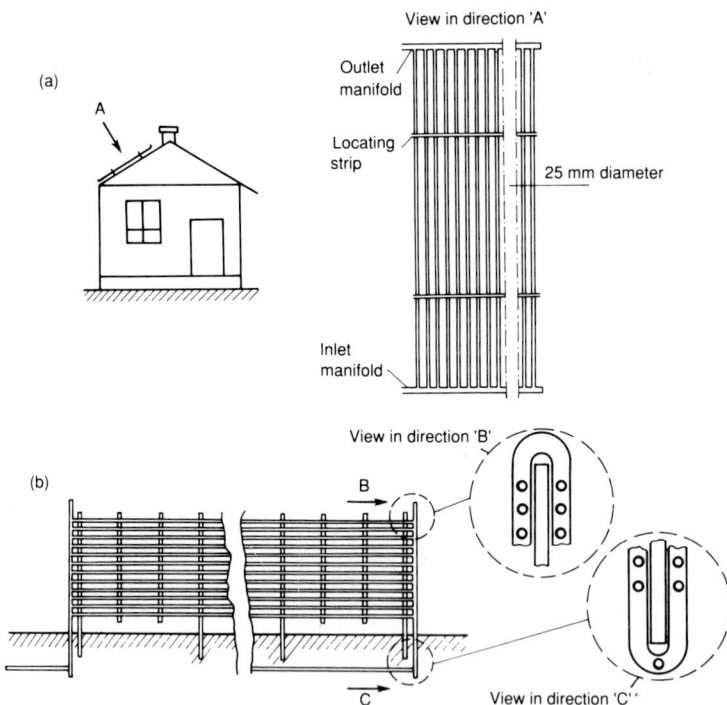

Figure 3.4 *The intake of energy from air constituting the low-temperature heat source. Heat exchangers air–brine (glycol): (a) tube bank type, located on the roof of a building; (b) fence type construction, made from finned tubes with natural convection of air; (c) in a sheet metal enclosure; (d) no enclosure, the exchanger is formed as a tower; (e) evaporators made of lamellae tubes, cooled directly with air, compressor located in the evaporator enclosure*

(c) Discharge manifold

(d)

(e)

Compressor

Compressor

20 mm diameter

ø10

Deflector

A

A

A-A

Compressor

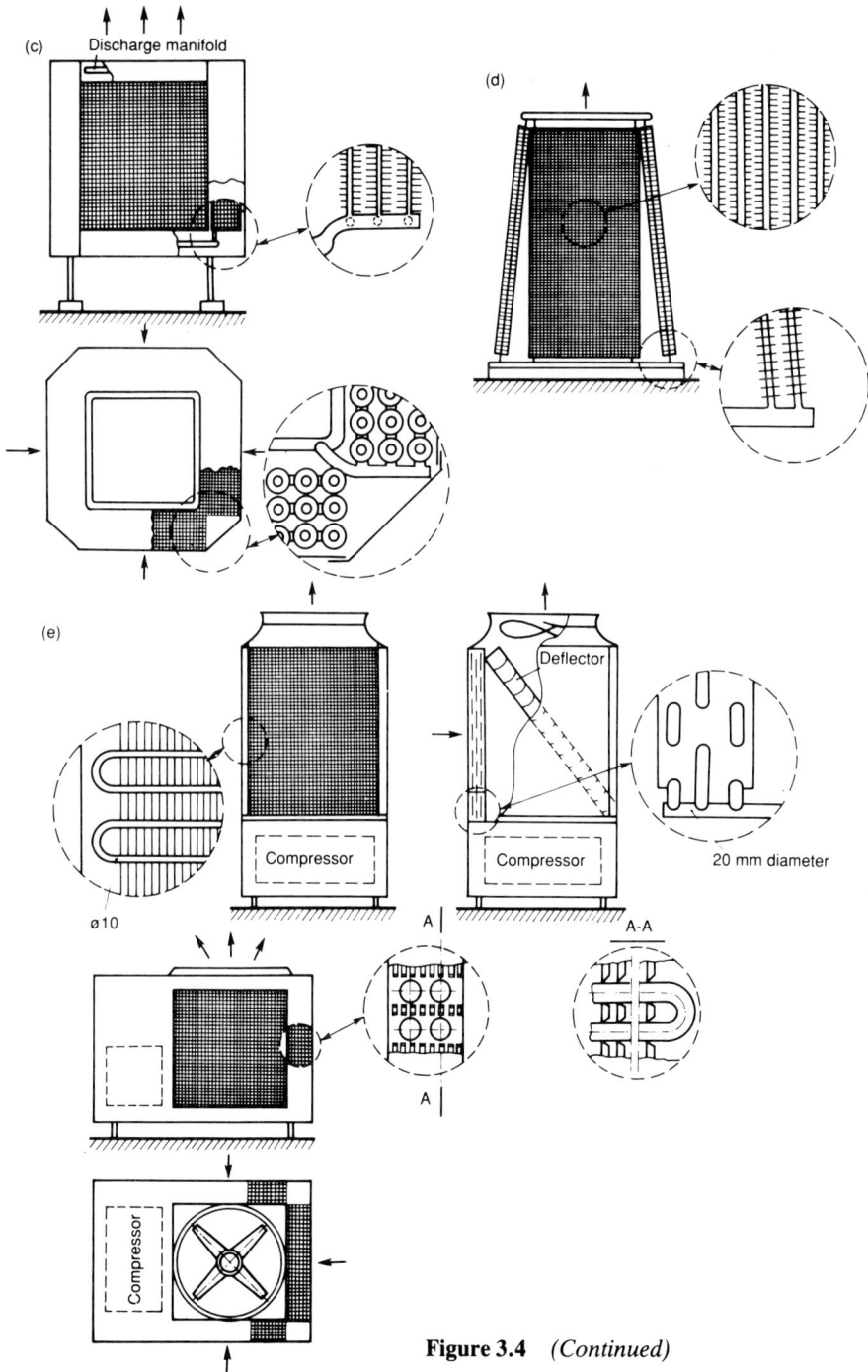

Figure 3.4 *(Continued)*

high-temperature heat source level is increasing, icing occurs at the heat exchanger. Icing reduces the heat transfer intensity, which manifests itself as a decrease of coherence of the low-temperature heat source. The heat exchanger, which is an intake of the low-temperature heat source in the form of atmospheric air, can also act directly as the heat pump evaporator (Figure 3.4e). There are, however, some arguments for using an intermediate heat exchanger employing an additional liquid medium (glycol or brine), despite the accompanying increase of exergetic losses. One of these is a tendency to build an inexpensive installation with a non-pressurized heat exchanger (gauge pressure at the evaporator is positive, e.g. for the R12 working fluid at 283 K it is 0.424 MPa, at 263 K it is 0.219 MPa). The second argument is the compactness of the evaporator, which ensures its required gas-tightness. The source intake heat exchanger often has to be located at a considerable distance away from the heat pump. To make the air flow independent of the wind direction and to ensure its good flow from all sides, the heat exchanger should be located on the roof or at some distance from the building. If one cannot install the compressor somewhere close, it is better to use an intermediate heat exchanger.

3.3.2 Surface waters

Surface waters can be used as the low-temperature heat source for large heat pumps. The energy reserves of larger rivers and lakes are utilized.

Table 3.1   *Examples of energy reserves in rivers with their waters used as the low-temperature heat sources*

River and location	Flow rate (m³/s)		Energy flux (MW)[a]		Page number in [34]
	Annual average	Minimum	Annual average	Minimum	
Wisla	624		3012		223
Nadwilanowka		320		1340	
Wisla	1.64		6.9		218
Wisla		0.09		0.38	
Rabka	0.9		3.8		240
Rabka		0.17		0.71	
Bzura	3.44		14.5		316
Leczycaq		0.84		3.53	
Utrata	2.59		10.9		319
Utrata		0.63		2.65	
Krutynia	0.57		2.4		301
Borowski Las		0.11		0.46	

[a]Assuming temperature drop of 1 K.

The energy in surface waters originates from the heat exchange between the atmospheric air and the soil. Even the small flow rate rivers carry relatively large reserves of energy, illustrated in Table 3.1 [34]. The table lists the energy fluxes, based on the assumed single stage temperature drop. If this energy is to be absorbed as heat, one can for example, intake 1/5 of the total river flow and have a source temperature drop of 5 K. As the temperature of water in the rivers is determined by heat transfer with the environment, the energy intake can be accomplished many times along the course of the river flow. An example of the temperature distribution in a river is shown in Figure 3.5a.

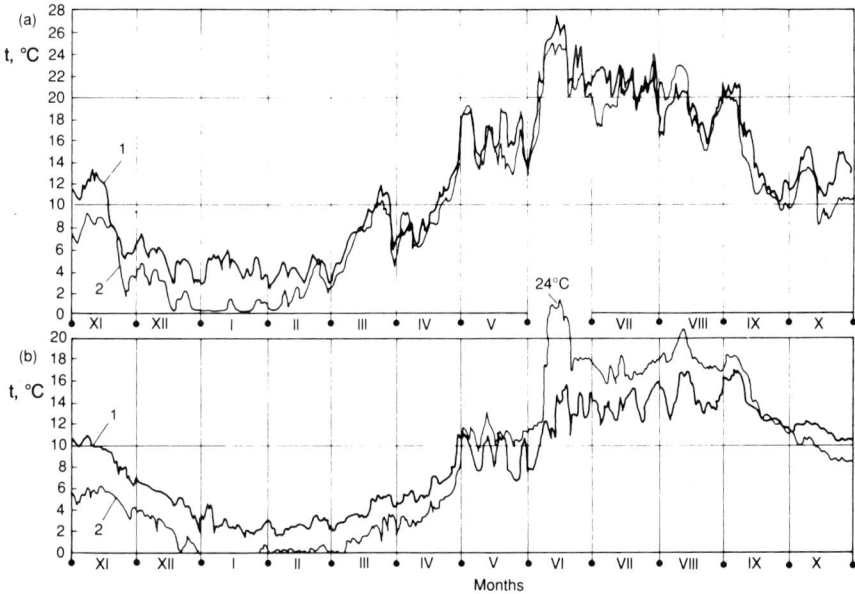

Figure 3.5 *Examples of annual temperature distributions of surface waters.* (a) *Rivers: 1, Vistula (Cracow, Bielany district); 2, Vistula (Warsaw, Nadwilanowka district).* (b) *Lakes: 1, Mamry (Przystan location); 2, Myczkowice (Myczkowo location)*

Surface waters as a low-temperature heat source have disadvantages, namely problems with energy intake during the periods of low-temperatures and at minimum flows, and also the icing of heat exchangers occurring at temperatures close to °C. Because of the high construction costs of the intake and because of the above limitations, river water should be used as the low-temperature heat source in large heat pumps (e.g. for heating municipal water) and during the periods when its temperature is higher (spring, summer).

The diagrams of surface water intakes, which can be used as the low-temperature heat sources for heat pumps, are shown in Figure 3.6.

(a)

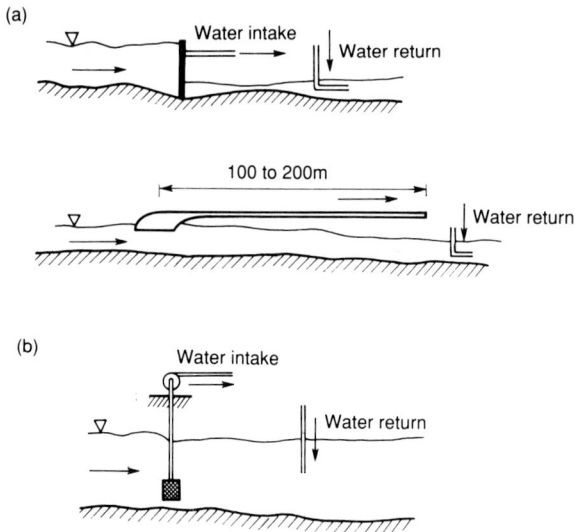

(b)

Figure 3.6 *Diagrams of intakes of the surface waters constituting the low-temperature heat source. Intake of the river water without:* (a) *using a pump;* (b) *with a pump*

3.3.3 *Underground waters*

Underground waters can be used as the low-temperature heat source in small heat pumps. As these waters absorb thermal energy directly from the soil, and indirectly from the sun, they are a source of good coherence and easy accessibility. Exemplary annual temperature distributions of underground waters (after [35]) are shown in Figure 3.7. Similar conditions prevail with utilization of waters from deep wells which exhibit small temperature variations throughout the year (278–285 K). The diagrams of intakes for underground waters are shown in Figure 3.8. These waters can be directed straight to the evaporator, and in cases of high salinity an intermediate heat exchanger can be used. Water from natural springs can only be considered theoretically, due to its low accessibility and limited energy reserves. Figure 3.7b illustrates this problem to some extent.

Utilization of thermal waters, heated by geothermal energy, from natural springs or deep boreholes, is a separate issue. In Poland's geological conditions these possibilities are quite limited; about 100 sources have been investigated, of which approximately 50 are utilized. The largest of these are listed in Table 3.2, and the more important ones, according to [35], have been charted on the map (Figure 3.9), which also gives the depths corresponding to temperature 323 K and shows areas particularly attractive for the utilization of geothermal energy. The existing sources are presently used for physiotherapeutic (balneological) purposes; hence the research was oriented towards such use, and thus refers mainly to waters of relatively high temperature. For heat pump use, the temperatures of these waters

Table 3.2 *Largest naturally discharging thermal springs utilized in Poland*

Name of borehole	Depth (m)	Temperature (°C)	Flow rate (m³/h)	Energy flux (MW)
Zakopane IG-4	1550–1560	39.0	170.0	7.7
Ladek 2	700	45.5	113.0	6.0
Koto IG-3	1773–1796	59.0	80.0	5.5
Cieplice 2	750	63.3	43.8	3.2
Sroda IG-2	1012–1020	40.0	40.0	1.9

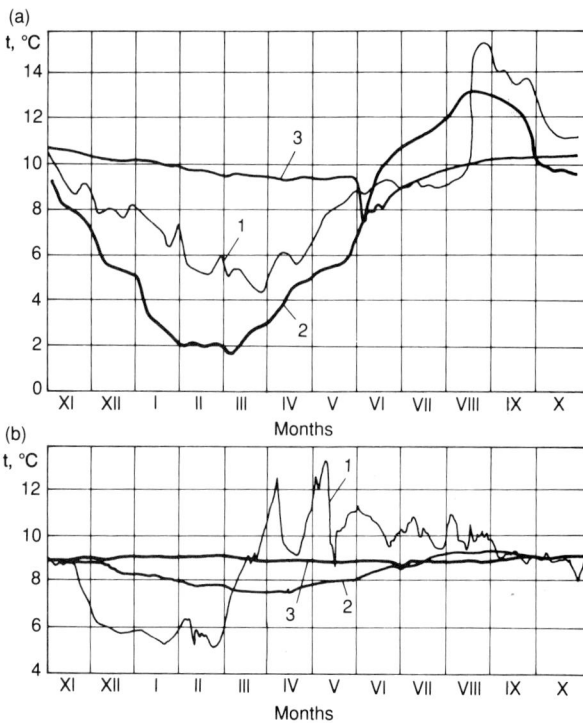

Figure 3.7 *Examples of annual temperature distributions of underground waters.*
(a) *Underground waters (in wells): 1, around Lowicz in the Bzura river basin, typical seasonal changes in Poland; 2, Kramrzyny location in the Wieprz river basin, lowest winter temperatures; 3, Brzeznik location in the Bobr river basin, little variations of temperature.* (b) *Springs: 1, Nowy Ludwikow in the Pilica river basin, large temperature variations, large spring 0.041–0.116 m³/s; 2, Kromolow location in the Warta river basin, average temperature conditions, middle-sized spring, 0.004–0.0017 m³/s; 3, Sulbiny in the Wilga river basin, small spring with a nearly constant outflow, 0.00115 m³/s*

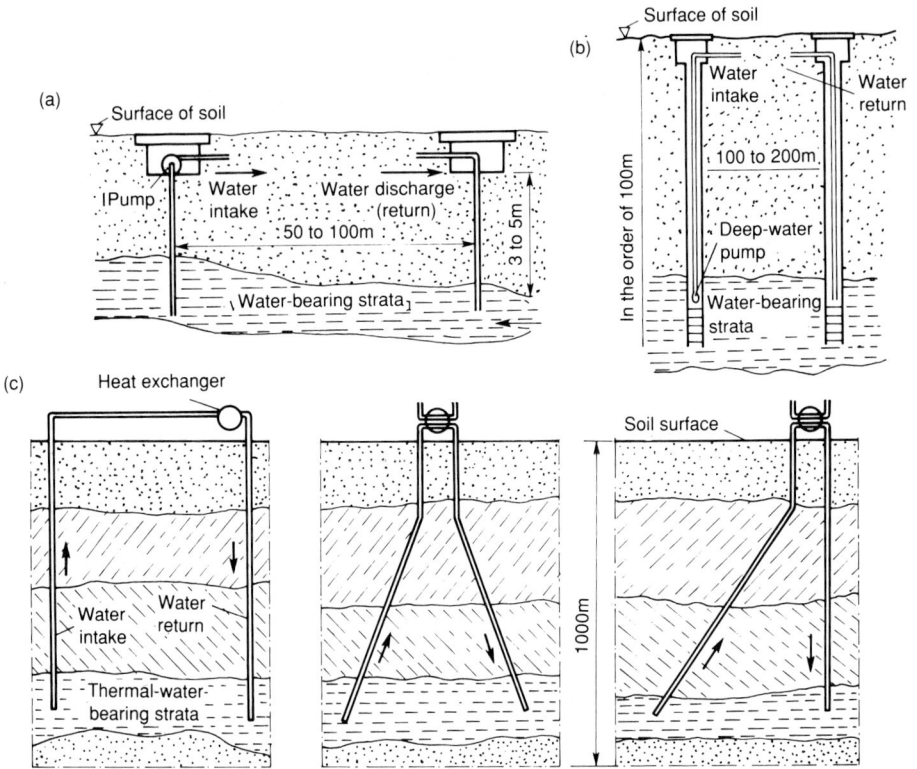

Figure 3.8 *Diagrams of the intakes of underground waters constituting the low-temperature heat sources:* (a) *surface;* (b) *drilled boreholes;* (c) *thermal*

can be lower and the wells can be shallower and, thus lower costs of intakes would follow. A rational solution, apart from a borehole for the water intake, requires another for the discharge, reaching the same geological strata. The borehole for water discharge must be situated at a sufficient distance so that water can be heated during its flow through the porous rocks (Figure 3.8c). Examples of applications of geothermal sources are given in [37].

3.3.4 Soil

The soil can be used as a low-temperature heat source only for small heat pumps. Energy collected from the soil originates from solar energy and from heat exchange with the atmosphere; it is accumulated in the top layer of soil, approximately 10 m deep. At this depth the soil temperature is equal to the yearly average air temperature (in our conditions this equals approximately 283 K). The best solution would be to collect the energy from this level, but because of high costs, such

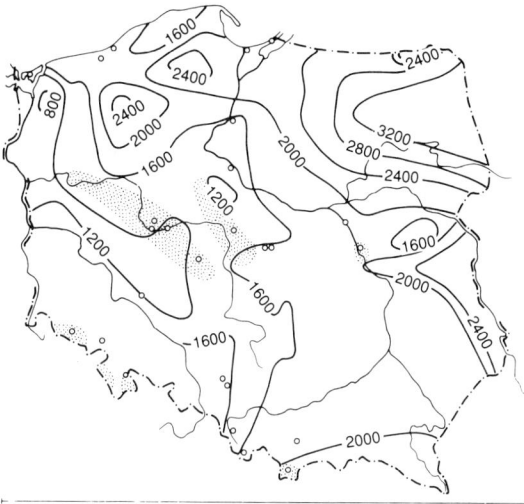

Figure 3.9 *Locations of thermal waters in Poland: numbers denote the isolines of depth (in metres) of waters at temperature 323 K. ○, existing boreholes and springs with thermal waters; shaded areas, areas potentially capable of yielding waters at temperatures over 300 K and with a spring yield over 20 m³/h*

solutions are not used. Energy is usually collected from a depth of 1–1.5 m, and at this depth the temperature varies throughout the year, approximately following a sinusoidal curve (in Poland's conditions, between 290 K in July and 278 K in January).

The low-temperature heat source heat exchanger is usually constructed from a tube formed into a flat coil (Figure 3.10). Glycol or brine flows inside the pipe,

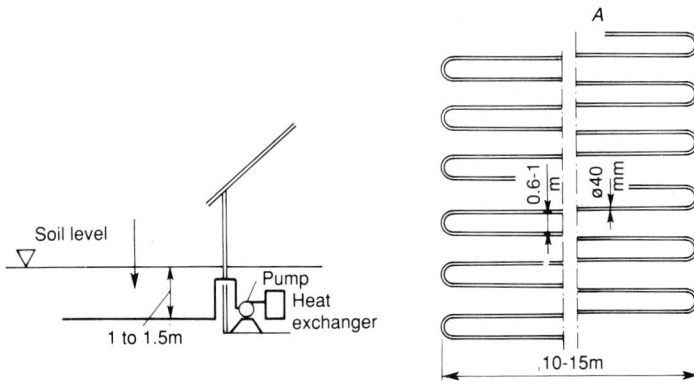

Figure 3.10 *Diagram of arrangement for the soil to act as a low-temperature heat source; exchanger tubes are buried at a depth 1–1.5 m*

absorbing heat from the soil and passing it on to the evaporator. At this depth the soil does not normally freeze, although the removal of heat can cause local freezing, which does not impair the functioning of the heat exchanger, and even slightly intensifies the heat transfer process. The intensity of heat transfer on the soil side is low. An investigation of the conditions of heat transfer was performed [38] on the basis of a solution of the heat conduction equation with boundary conditions taking into account the geometry and thermal conductivities of the heat exchanger (Figures 3.11 and 3.12). Here we shall limit ourselves to an assessment

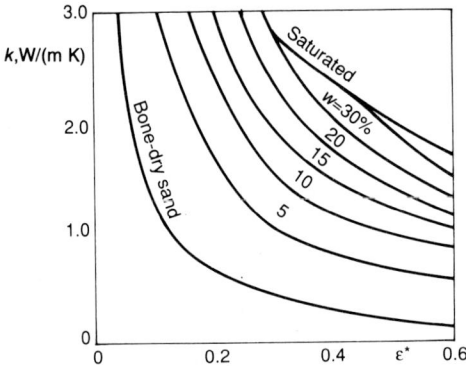

Figure 3.11   *Thermal conductivity, k, of silica sand as a function of porosity and relative humidity defined as the percentage mass fraction of water*

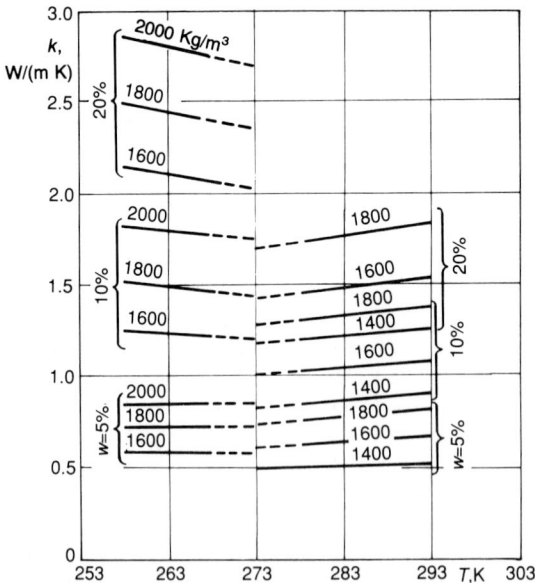

Figure 3.12   *Thermal conductivity, k, of clay as a function of temperature, humidity and density. Solid lines, experimental data; dashed lines, extrapolated data*

of these conditions. With a temperature difference of 5–10 K one obtains a heat flux in the region of 15–50 W/m per 1 m of length of 30-mm-diameter pipe. After taking into account the necessary pitch of pipe rows, each 1 kW of heat flux absorbed in the evaporator requires 12–100 m² area of soil, so this area is relatively large. For this reason the soil can be used as the low-temperature heat source only for small heat pumps, installed in sparsely built-up areas, not otherwise utilized and with no trees, since the process of installing a heat exchanger requires extensive earthworks. This problem is discussed in [39] for conditions in Sweden.

3.3.5 Solar energy

The sun radiates out an energy flux of 4×10^{23} kW, of which only 1.9×10^{14} kW reaches the earth's atmosphere. This is actually 10 000 times more than the present world energy requirements. The surface density of this flux at the atmosphere's boundary is equal to 1.38 kW/m², on the earth's surface at moderate latitudes it drops to 0.35–1.0 kW/m². This energy reaches the earth's surface as either direct or scattered (indirect) radiation. Approximately 50% of this energy arrives as scattered radiation (in Polish conditions, in summer, in the morning up to 70%, at noon 40%; in winter, in the morning 90%, at noon 65%). The radiation scatters during its travel through the atmosphere, meeting on its way the atoms of gases, dust, droplets of water, etc. The longer the path of radiation through the atmosphere, the larger the degree of scatter (in the morning and evening it is longer than at noon, in winter larger than in summer). Energy reaches the earth's surface as scattered radiation even on overcast days, when the direct radiation energy does not get through. The maximum utilization of direct radiation occurs when the surface receiving the radiation is at right angles to the direction of the rays, while for scattered radiation the surface receiving the radiation should be horizontal. If this surface is the surface of a collector, one needs to orient it in such a way that both kinds of radiation are utilized to a maximum extent. Good utilization of direct radiation is the more difficult of the two problems. The orientation of a collector and possibly its motion following the sun's relative daily travels requires a thorough analysis. It is important at latitudes having larger intensities of solar radiation. The surface density of energy flux of direct and indirect radiation for the city of Warsaw during different months of the year is shown, after [40], in Figure 3.13. These data give us some insight into the power of solar radiation. The power is not, however, a good indicator, as it can vary during the day depending on climatic conditions (cloud cover, rain, etc.). A better indicator is provided by the energy which can be accumulated during a day or a month, taking into account the statistically averaged time of cloudless weather (Figure 3.14). Large variations of this energy can occur each year, up to 100% in individual months; still larger variations may be present on a daily basis, both within one day and from one day to another. For this reason a low-temperature heat source utilizing solar energy requires installation of an additional device for the accumulation of energy (e.g. a water tank). This increases the already high costs of the source. One can, however, by using such sources obtain, in favourable external conditions, temperatures sufficiently high for the

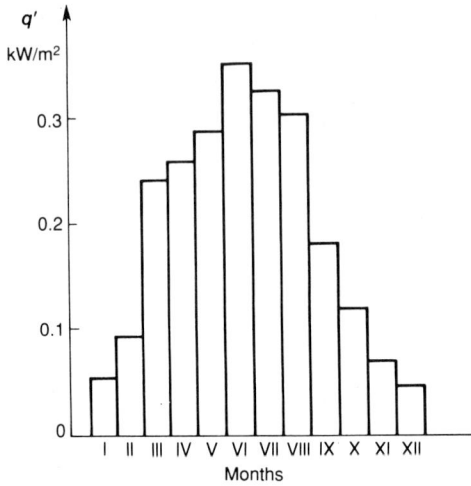

Figure 3.13   *Surface density of the energy flux of total solar radiation for Warsaw in individual months of the year (average data for the period 1961–1978)*

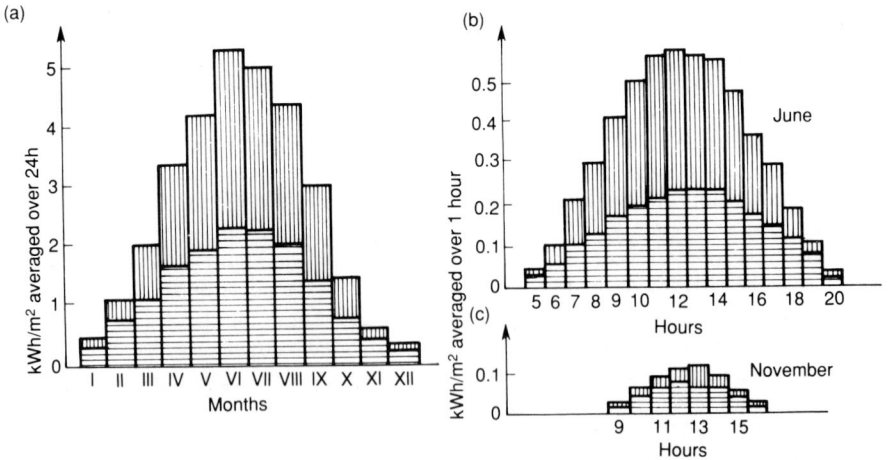

Figure 3.14   *Average values of the surface density of energy flux for direct (vertical hatching) and indirect (horizontal hatching) radiation, per 1 m² surface area in Warsaw: (a) averaged over 24 h, during various months of the year; (b) averaged over 1 h, during a selected day (average values for years 1961–1978)*

user in water circulating in the collector. The heat pump can be employed to raise this temperature further still, or under less favourable external conditions.

The low-temperature heat source intake is provided by a collector absorbing the radiation energy and passing it on to water (or water with glycol) circulating in a closed circuit at temperatures between 303 K and 313 K, perhaps even higher. The intake shown in Figures 3.4a and 3.4b partially utilizes the radiation energy (heat exchanger situated on the roof of a building). For a year-round operation more advantageous are collectors covered with acrylic sheets (Figure 3.15a,b); one can also use absorptive panels made of copper sheet and fitted underneath with pipes for water flow. To increase the absorptivity, the side exposed to radiation is usually covered with black paint or with special selective paints (black nickel, black chromium), for which the emissivity within infrared range is smaller than absorptivity, leading to a substantial increase in the efficiency of absorption. The designs of collectors for collecting mainly the direct radiation are shown in Figure 3.15c.

The densities of heat flux, which can be obtained from the surface of a collector, range from 0.12 to $0.3\,kW/m^2$, sometimes up to $0.5\,kW/m^2$. It means that for a heat pump with the power of the low-temperature heat source 15 kW, the necessary surface area of a collector built from panel elements (Figure 3.15c) is several dozen square metres.

3.4 Artificial low-temperature heat sources

The artificial sources utilize waste heat from technological processes of various industries, conventional and industrial energetics, municipal engineering, etc. Because of their individual characteristics, one needs to analyse all the conditions relating to the use of these sources, beginning with the kind of substance in the source, its phase, aggressiveness, and suitability for heat transfer operations and flow. The substance can be in gas phase (air or vapours); more often it is a liquid, usually water of different clarity (from clean water to sewage). With aggressive liquids one needs to remember corrosion and deposition of sediments on the exchange surface. The analysis of a source should also include its parameters, power, temperature range and variations of temperature (within a day, week and year).

Sources of larger power (from several hundred kilowatts to several megawatts) are more suitable for use with a heat pump. As yet no heat pumps with power larger than a dozen or so megawatts have been installed, although industry, particularly energetics, offers sources of waste heat of much larger power. The temperatures of artificial sources are usually higher than those of natural sources (with the exception of thermal waters). This creates very attractive conditions for the use of heat pumps. The changes of parameters within a year define the possible thermal output, and the most advantageous conditions exist when the source is available during a significant part of the year (e.g. 5000 hours) and meets the energy requirement. Artificial sources exist at the location of processes which produce waste heat as a byproduct. If it is to be utilized in heat pumps, there must be a requirement for heat of higher exergy. Usually there are no problems with

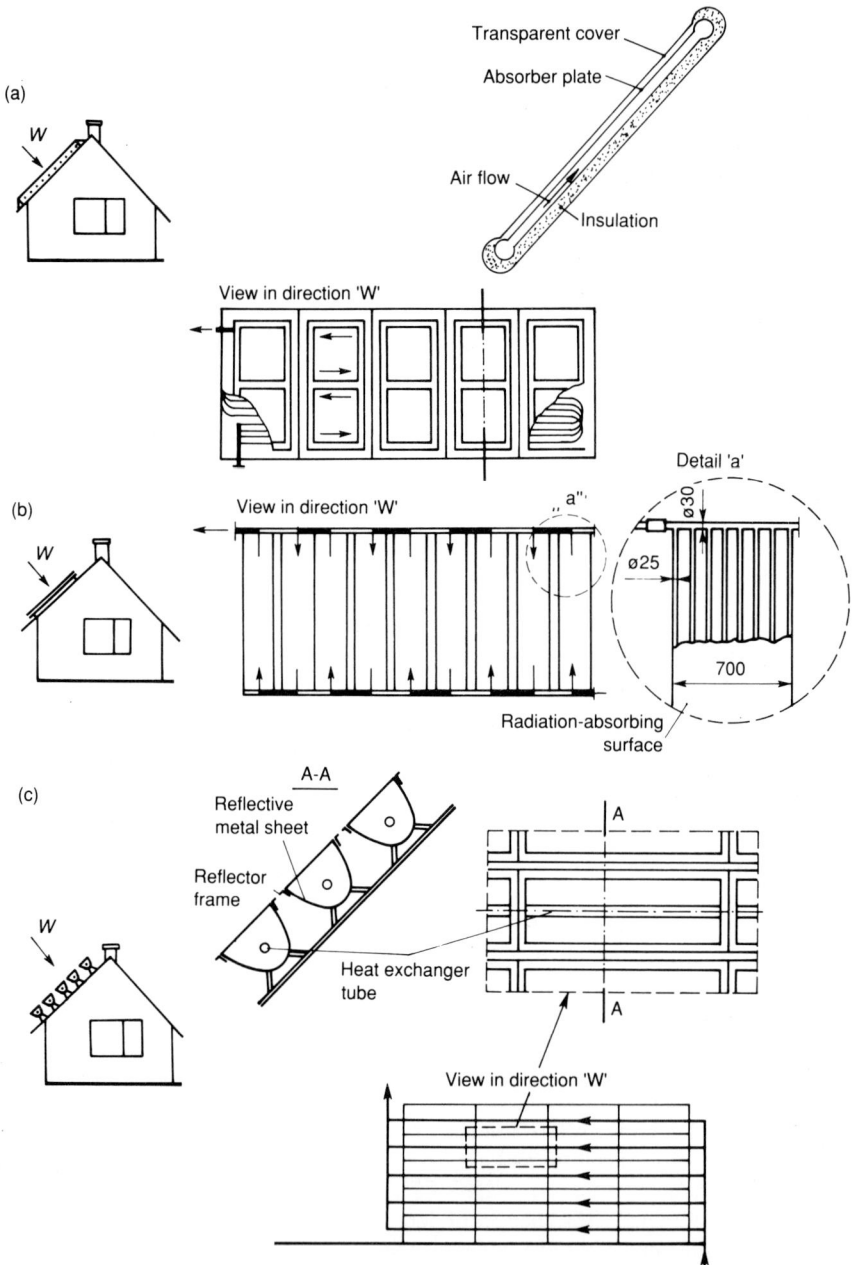

Figure 3.15 *Solar collectors (intake of the low-temperature heat source energy in the form of solar radiation) for heat pumps used in domestic heating: (a) heat exchanger covered with glass; (b) collector with a copper sheet absorption surface; (c) collector with reflectors, designed to absorb mainly direct radiation*

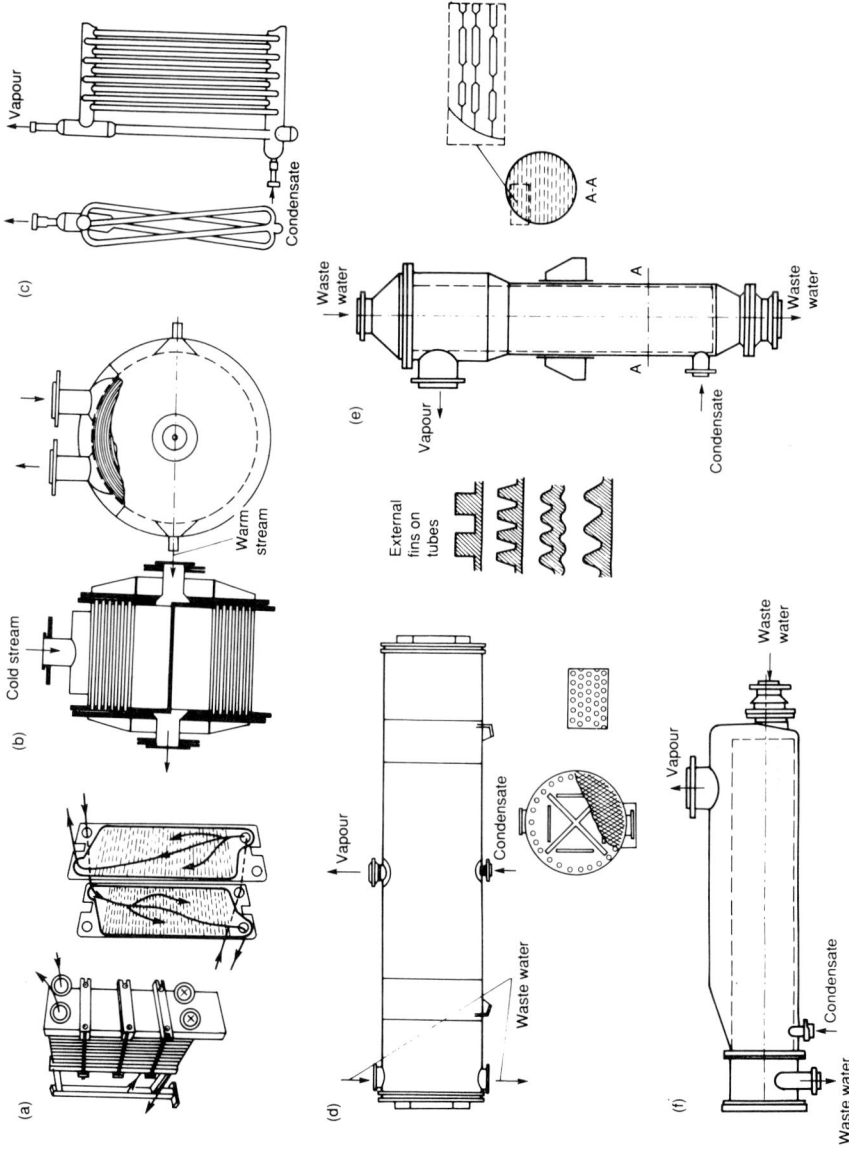

Figure 3.16 *Examples of intakes for the artificial low-temperature heat source in the form of waste heat in water: intermediate heat exchangers (water–water): (a) plate (up to 2000 m²); (b) spiral (to 200 m²). Evaporators directly fed with water from the low-temperature heat source: (c) steep tube bank (up to 50 m²) directly immersed, with forced flow; (d) shell and tube (up to 300 m²). Lamellae: (e) vertical; (f) horizontal (up to 800 m²)*

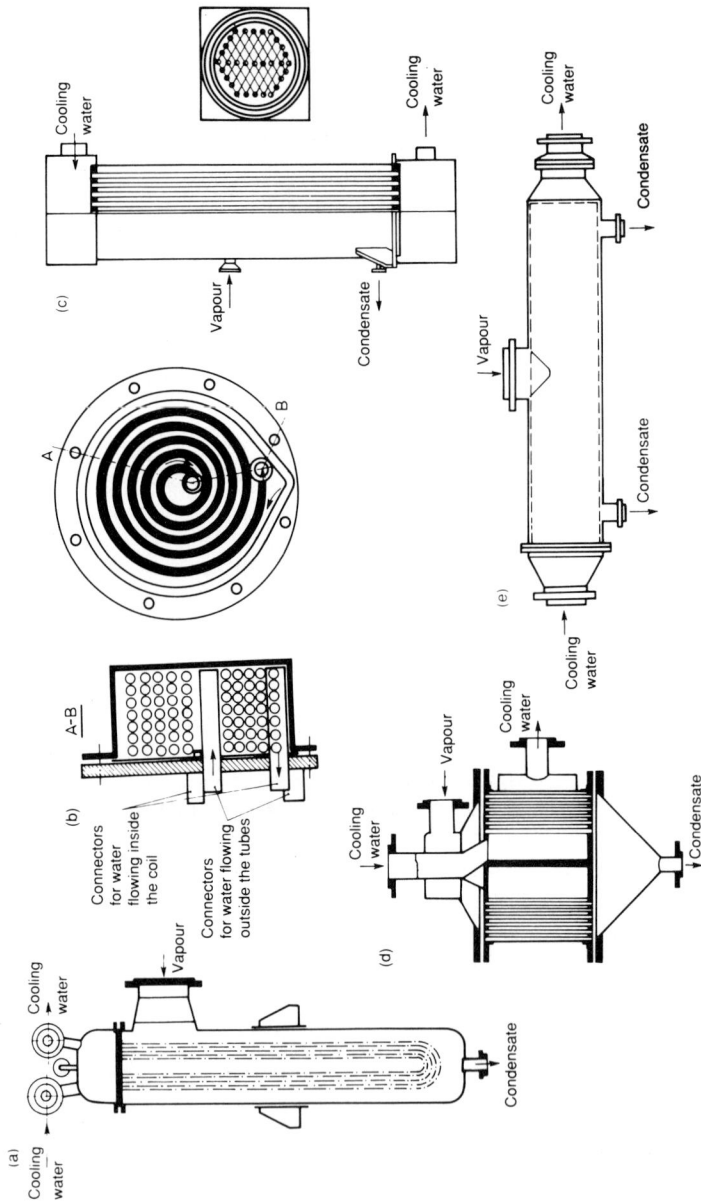

Figure 3.17 *Heat exchangers which can be used to transfer heat at the high-temperature heat source level. For small heat pumps up to 30 m²: (a) with vertical tubes; (b) coiled. Condensers for large heat pumps: (c) vertical shell and tube (up to 250 m²); (d) spiral (up to 200 m²); (e) lamellae (to 800 m²)*

such requirements; most often heat at the high-temperature heat source level is returned to the process from which it was collected at the low-temperature heat source level. It can also be utilized externally.

The low-temperature heat source intake can be realized in several ways. The substance can flow directly through a heat exchanger (evaporator or desorber), which then constitutes its intake, or the heat transfer between this substance and the working fluid can be realized by using an intermediate fluid. This requires a second exchanger, in which heat from the source substance is transmitted through the intermediate fluid to the low-temperature heat source. The flow of a source substance is either forced by using a pump, or the heat transfer surface is immersed in the source. The designs of heat exchangers collecting the energy from the low-temperature heat source are shown in Figure 3.16. Among these are evaporators, additional heat exchangers and desorber exchangers. The heat transfer surface areas in these exchangers (evaporators) can range from several dozen to several hundred square metres, depending on the collected power and design type. Should larger powers be necessary, the exchangers can be set into batteries. Correct calculations of the heat transfer surface lead to a reduction of exergetic losses within the low-temperature heat source. The calculation methods currently being developed allow optimization of the evaporator design.

3.5 The high-temperature heat source

The high-temperature heat source is a condenser (absorber) cooled by a substance, usually water. This substance carries the energy from the low-temperature heat source to the place of its utilization. Flow of water through the condenser is usually forced by a pump, although there are also condensers with naturally induced flow. The water used is usually clean, which makes easier both the technical implementation of the exchange process and the calculations. Calculation methods applied to condensers enable their rational design and optimization and maintenance of the exergetic losses at an adequate level. Many detailed data on the high-temperature heat sources are given in [41]. Typical designs of condensers are shown in Figure 3.17.

Chapter 4

Mechanical vapour compression heat pumps

4.1 Methods of describing the cycle of mechanical vapour compression heat pumps

As already mentioned in Chapter 1, to facilitate the description of practical cycles, one can use the comparative cycles. These are theoretical cycles with varying degrees of complexity, from simple to well developed. They can be regarded as consecutive approximations of the practically implemented cycle. In a simply described cycle, the matrix of parameters defining the cycle has few terms, and the parameters can be estimated without difficulty. This results from the fact that such a cycle consists either of processes which are all reversible (ideal cycle), or it allows one or two irreversible processes (theoretical cycle), simple in their description, i.e. with an easy-to-estimate loss accompanying this irreversibility (the loss can be expressed either as an entropy increase, or as a work loss, or as an exergy loss). As the consecutive descriptions are formed, ever more irreversible processes and thus possible losses are introduced (approximate cycle), and in the most developed description all the processes are irreversible. The matrix of parameters describing the cycle is then quite large. Estimating the individual parameters becomes more difficult as well. Particular difficulties arise when the description of losses must take into account the effects of elements of the plant in which these processes are implemented and where the arising losses are defined by the structural quality of the elements.

An important characteristic of every theoretical cycle is its ability to emulate a practical cycle; hence the necessity to compare the cycles. Although the main aim is to compare the theoretical cycle with a practical one, the possibility of comparing various theoretical cycles is also important. To this end one can compare the matrices of cycle parameters, obtaining the matrices of comparisons. It is more convenient to express such a comparison using just one numerical value (coefficient), but one then needs a more global way of comparing the cycle parameters. This is possible, but requires introducing certain disciplines for such comparisons. First of all, one can only compare the cycles which work within the same range of intensive parameters (T_{hs}, T_{ls}). Secondly, taking into account the primary function of a heat pump, one can assume that only those cycles will be compared which discharge the same flux of exergy to the high-temperature heat source. In such cases the two cycles can be compared using a coefficient defined by a ratio of driving exergies of these cycles. As in mechanical vapour compression heat pumps the driving exergy is usually in the form of work; instead of driving exergy, one can talk about work. Additionally, remembering that the real aim of the whole description procedure is to estimate the work of a practical cycle, which is always larger than the work of theoretical cycles (particularly those with a larger number of reversible processes), the coefficient is smaller than unity:

$$\text{cycle comparative coefficient} = \frac{\substack{\text{work delivered to a cycle m, which} \\ \text{takes into account fewer losses} \\ \text{(irreversible processes)}}}{\substack{\text{work delivered to a cycle n, having} \\ \text{larger number of losses}}} \qquad (4.1)$$

$$\varphi_{nm} = w_m/w_n \leqslant 1$$

As there may be many comparative cycles, there can also be many coefficients obtained from this definition. Here are some examples:

$$\varphi_{r,iR} = W_{iR}/W_r \text{ and } \varphi_{r,c,C} = W_C/W_r; \; \varphi_{a,c,iR} = W_{iR}/W_a$$

Not all of these are equally useful in the description, so some are not used, but it is worth looking at two groups of these coefficients. The first group is where the cycle under consideration is compared with an ideal reverse Rankine cycle; in such a case the comparative coefficient is the same as the exergetic efficiency and the following relationships are in force:

$$\varphi_{r,iR} = \varphi_{r,C} = \eta^*_{rc}; \quad \varphi_{niR,iR} = \varphi_{niR,C} = \eta^*_{niR} \qquad (4.2)$$

The second group comprises coefficients whose values are close to unity, obtained from the comparison of a practical cycle with a theoretical cycle with a well-developed description (called here an approximate cycle). Thus the following relationships occur:

$$(\varphi_{r,a} \approx 1) > \varphi_{r,niR} > \varphi_{r,iR} = \varphi_{r,C} \qquad (4.3)$$

The comparative cycles used in the description of a practical mechanical compression heat pump are shown in Figure 4.1.

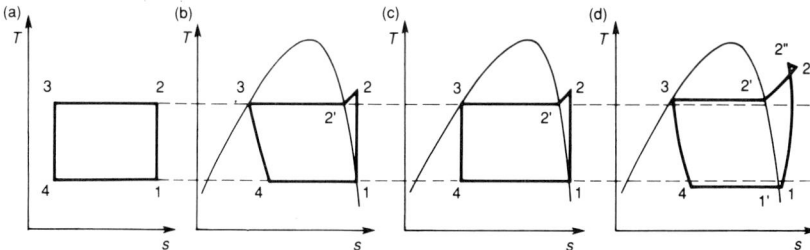

Figure 4.1 *Comparative (theoretical) cycles of a mechanical vapour compression heat pump:* (a) *Carnot;* (b) *non-ideal reverse Rankine cycle;* (c) *ideal reverse Rankine cycle;* (d) *approximate cycle*

The 'non-ideal reverse Rankine cycle' and the 'ideal reverse Rankine cycle' are sometimes referred to without distinction as 'reverse Rankine cycle' or even 'Rankine cycle'. While this simplification is adequate for general engineering usage, in the context of this book it is important to be able to distinguish between the irreversible nature of the non-ideal cycle and reversible ideal cycle. In Continental literature the 'non-ideal reverse Rankine cycle' is often referred to as the 'Linde cycle' which makes the definition uniquely precise, particularly as the true Rankine cycle, usually applied to the analysis of steam turbine cycles, employs compression of liquid, so the true reverse Rankine cycle should use an expansion engine.

The simplest cycle is the Carnot cycle. Often it is enough to know just two intensive parameters: the temperatures of the high- and low-temperature heat sources. Should there be a need to estimate the extensive parameters (heat fluxes), a single parameter describes the cycle: Δs. However, as the high-temperature heat source temperature in the Carnot cycle is constant, it cannot deliver in the high-temperature heat source, at the same temperature conditions, the same heat flux as the practical cycle. To comply with these conditions, one should use not one, but a whole series, of Carnot cycles, as illustrated in Figure 4.2. To avoid this procedure in the description, one more ideal reverse Rankine cycle is introduced, somewhat descended from the non-ideal reverse Rankine cycle, but having a reversible expansion process, instead of the irreversible throttling process which occurs in the non-ideal reverse Rankine cycle.

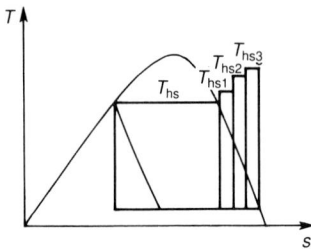

Figure 4.2 *A series of Carnot cycles which could replace the non-isothermal heat transfer at the high temperature heat source of a non-ideal reverse Rankine cycle*

The non-ideal reverse Rankine cycle is a theoretical one. It takes into account both the properties of the working fluid and suggests the devices, or in other words the elements of plant, suitable for implementing the individual processes. Thus it is in a way defined specifically for the purposes of describing a cycle implemented in a mechanical compression heat pump or in mechanical compression cycle refrigeration equipment. It has a sufficiently simple description, discussed in detail in Section 4.2.2. The remaining task of defining the comparative coefficient can be solved in two ways: the coefficient can be estimated experimentally or by theoretical calculation.

The experimental method relies on a direct measurement of the coefficient on a heat pump actually built. This method is favoured by the manufacturers, who can increase this coefficient in the consecutive versions of their design. Despite the significant effectiveness of this method, it has an important drawback: it considers all the losses globally. For this reason more general and thus better results can be delivered by the theoretical approach: isolating individual losses and proposing methods to describe them. Often these methods are not too precise, requiring the postulation not only of various conditions, but also of values of certain parameters, and often one does not have sufficient confidence in the assumptions made regarding these conditions and parameters. Because of this, such an approach can lead to even less precision than the experimental method. On the other hand, it explains the mechanisms of origins of the losses, thus giving an opportunity to influence their occurrence, and consequently their minimization. The approximate cycle, enabling a description and then superposition of all losses, is discussed in section 4.3. Since the description of losses obtained from theoretical discussion cannot be generalized to give, for example, a general formula for the cycle comparative coefficient, apart from the general discussion this section also gives a numerical example. Although all efforts were made to select representative numerical values, the results should be viewed as an illustration of the method rather than as a collection of numerical data, which could be applied to other cases of cycles of mechanical vapour compression heat pumps.

4.2 The non-ideal reverse Rankine cycle

The non-ideal reverse Rankine cycle comprises reversible processes except for the irreversible process of throttling (isenthalpic expansion) and is realized using a

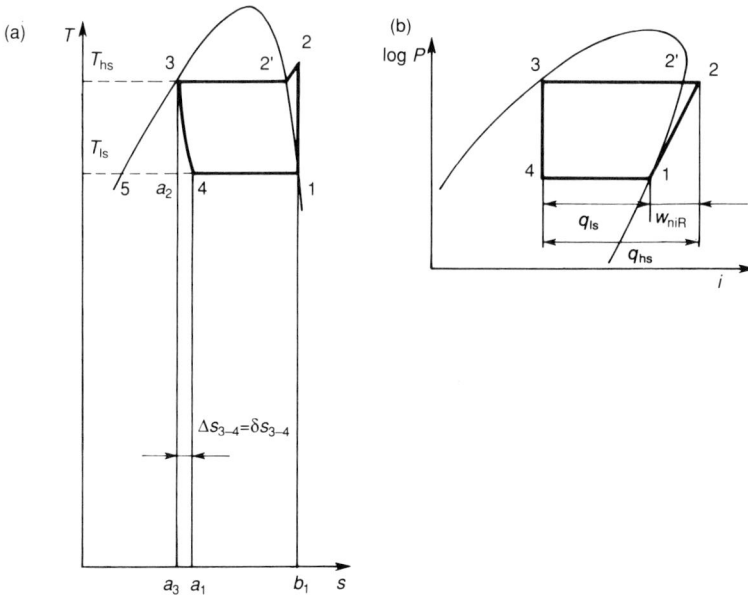

Figure 4.3 *The theoretical non-ideal reverse Rankine cycle: (a) in T–s coordinates; (b) in log P–i coordinates*

single-component, two-phase working fluid. The characteristic parameters (of the cycle) thus depend on the kind of fluid employed. The cycle is shown in Figure 4.3. The state of vapour before compression corresponds to point 1 (point 1 is located on the dry saturated vapour line), and the state after the isentropic compression corresponds to point 2.

The state of the liquid at the condenser exit is defined by the isobar 2–3 (point 3 is located on the boiling liquid line), and the state of the vapour at the evaporator exit is defined by the isobar 4–1. Thus both the heat discharge (process 2–3) and heat delivery (process 4–1) are isothermal and isobaric, the compression is isentropic (1–2), and the isenthalpic expansion (3–4) is an irreversible process. The type of such processes simultaneously defines the equipment needed to implement the cycle (compressor, condenser, throttling valve and evaporator).

Qualitative analysis of the non-ideal reverse Rankine cycle is conveniently done in the thermodynamic coordinate system T–s, in which heat and work are represented by the corresponding areas (Figure 4.3a): heat absorbed from the low-temperature heat source q_{ls}: by the area $b_1 14 a_1 b_1$; heat delivered to the high-temperature heat source q_{hs} by the area $b_1 122'3 a_2 a_3 a_1 b_1$; and work by the area $122'3 a_2 a_3 a_1 41$ or the equivalent area $122'35 a_2 41$. The T–s system is, however, of little use in direct calculations, because estimating the relevant values (areas on the chart) requires the use of a planimeter. For this reason it is more convenient to use such a system of thermodynamic coordinates, which allows direct readout of the quantities of heat and work. In a non-ideal reverse Rankine cycle heat is delivered and discharged at constant pressures, thus these quantities are defined by the enthalpy differences $q_{ls} = i_1 - i_4$, $q_{hs} = i_2 - i_3$, and work is also defined by the enthalpy difference $w = i_2 - i_1$, which is proved later in this section. It is therefore convenient in direct calculations to use a chart in which one of the coordinates is the enthalpy of the working fluid. The other coordinate might be pressure, shown in a logarithmic scale, so that the isentropic compression process is shown on this chart as a straight line (the same applies to a polytropic process, although it is no longer a non-ideal reverse Rankine cycle). Hence the i–$\log P$ chart is usually employed in the calculations of non-ideal reverse Rankine cycles.

In an analysis of the qualitative characteristics of this cycle, however, a T–s chart is more convenient and thus this chart is mainly used here. As is known, the process of throttling in a valve is irreversible. The First Law of Thermodynamics:

$$dQ = dQ_{ext} + dQ_{int} = di + dW_{loss} \tag{4.4}$$

for the process of throttling, in view of the conditions $dQ_{ext} = 0$ and $di = 0$, simplifies to:

$$dQ = dQ_{int} = dW_{loss} \tag{4.5}$$

for the process 3–4:

$$q_{3-4} = q_{int\,3-4} = \delta W_{loss\,3-4} \tag{4.6}$$

so that all the heat is produced from the dissipated work (it is thus the heat of dissipation), hence:

$$q_{3-4} = \overline{T}\,\Delta s_{3-4} = \overline{T}\,\delta s_{3-4} \tag{4.7}$$

where $\overline{T}$ = average temperature of the process.

Equation (4.7) leads to an important relationship:

$$\delta s_{3-4} = \Delta s_{3-4} \tag{4.8}$$

which means that the increase in entropy of an irreversible process (δs_{3-4}) is in the throttling process equal to the actual change of entropy of the process (Δs_{3-4}). The lost work and exergy loss, on the other hand, are correspondingly equal to:

$$\delta w_{\text{loss}\,3-4} = \overline{T}\delta s_{3-4} \tag{4.9}$$

$$\delta b_{3-4} = \overline{T}_{\text{en}}\,\delta s_{3-4} \tag{4.10}$$

which is illustrated in Figure 4.3a, where the heat of the process, equal to the heat of dissipation and the lost work, is represented by the area $a_143a_2a_3a_1$.

It is easy to show that the lost work is equivalent to the area $435a_24$. To prove this statement, Figures 4.4c and 4.4d show enthalpy fields at points 3 and 4, with a

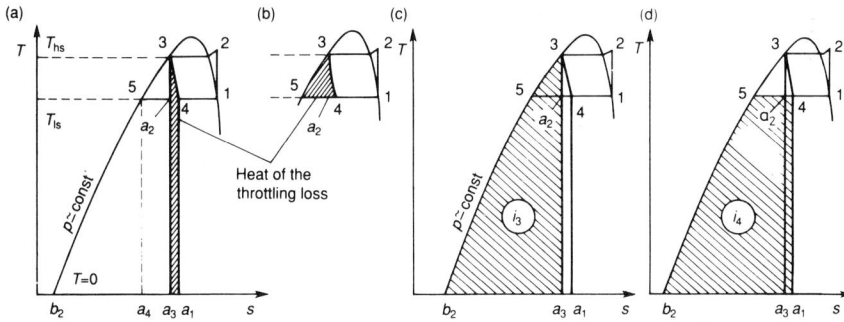

Figure 4.4   *Loss in the expansion valve of a non-ideal reverse Rankine cycle: (a) heat of the process is equal to the cross-hatched area; (b) the cross-hatched area is equivalent to the area in (a); (c) and (d) auxiliary diagrams proving the equality of areas in (a) and (b)*

simplifying assumption that the specific heat of the liquid does not depend on pressure (constant pressure lines coincide with the boiling liquid line). For the condition $i_3 = i_4$ the cross-hatched areas in Figures 4.4c and 4.4d are equal, so areas a_235a_2 (Figure 4.4c) and $a_14a_2a_3$ (Figure 4.4d) must also be equal, and hence the cross-hatched areas in Figures 4.4a and 4.4b are equal as well. The work of a non-ideal reverse Rankine cycle is thus defined by the area $1235a_24$. This results in a simple definition of the work of the cycle. Namely, taking into account that enthalpy at point 1 is equal to the cross-hatched area in Figure 4.5a, and at point 2 to the cross-hatched area in Figure 4.5b, so that the work, equivalent to area 123541, is equal to the difference of these two areas, and thus to the difference of enthalpies $i_2 - i_1$, on the i–log P chart (Figure 4.3b), this work is represented by the distance $w = i_2 - i_1$. It is also easy to show that $q_{\text{ls}} = i_1 - i_4$ is equal to the area $b_114a_1b_1$ in Figure 4.5a, and $q_{\text{hs}} = i_2 - i_3 = i_2 - i_4$ to the area $b_1123a_2a_3b_1$ in Figure 4.5a.

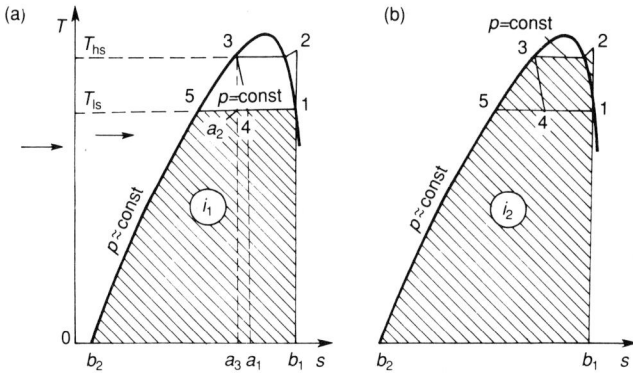

Figure 4.5   *Work of the non-ideal reverse Rankine cycle, equal to the difference of enthalpies:* (a) *enthalpy at point 1;* (b) *enthalpy at point 2*

Since the comparative coefficient for a practical cycle compared with the non-ideal reverse Rankine cycle is quite large ($\varphi_{r,niR} \sim 0.7$), analysis of the non-ideal reverse Rankine cycle yields quite a lot of information on the practically realized cycle. Such analysis can be conducted for several versions, because of the number of variables involved: selected working fluid, temperatures T_{hs} and T_{ls} and the COP_{niR}. It is worth remembering here that the minimum value of COP for a practically realized cycle of a mechanical vapour compression heat pump can be assumed to be not smaller than three. Only then will the energetic efficiency (referred to primary energy used to drive the heat pump) be larger than the efficiency obtainable from other heating systems. Thus COP_{niR} should not be smaller than four. Figure 4.6 shows the results of calculations for the non-ideal reverse Rankine cycle employing the R12 working fluid. These results give a good idea of the high-temperature heat source temperatures which can be obtained at a good energetic efficiency.

An important element determining the characteristics of a non-ideal reverse Rankine cycle is the kind of working fluid employed; Figure 4.7 shows non-ideal reverse Rankine cycles operating within the same temperature ranges, but using different working fluids, and Table 4.1 lists the characteristic parameters of these cycles. The results shown in Table 4.1 make it possible to compare the suitability of fluids under consideration for use in a non-ideal reverse Rankine cycle. This comparison is based on criteria listed in Table 4.2. Using a grading scale ranging from 4 (good) through 3.5 (quite good) to 3 (satisfactory), one can compile a list of quality of the analysed fluids, including the average score (Table 4.2).

So far in the description of non-ideal reverse Rankine cycles we have used the thermodynamic charts and diagrams, to some extent suggesting their use in calculations. Indeed, these charts are widely used. In numerical calculations, however, it is convenient to use analytical formulae to describe both the properties of fluid and parameters of the cycle.

Table 4.1 A listing of calculation results for the theoretical non-ideal reverse Rankine cycles, as illustrated in Figure 4.7, employing different working fluids, operating within the temperature range 273–343 K; for this temperature range $COP_C = 4.9$

No.	Working fluid	Enthalpy values (kJ/kg)			Heat $i_2 - i_3$ (kJ/kg)	Work $i_2 - i_1$ (kJ/kg)	Pressure (MPa)		Compression ratio	v (m³/kg)	COP_{niR}	η^*_{niR}	V^a_{rel}
		i_1	i_2	i_3			Condenser	Evaporator					
1	R22	304.00	348.75	192.60	156.15	44.75	2.990	0.50000	5.98	0.04710	3.49	0.71	1.000
2	R12	254.70	288.10	171.90	115.50	33.90	1.900	0.31000	6.10	0.05667	3.40	0.71	1.620
3	R134a	244.40	285.50	156.10	129.40	41.10	2.167	0.30000	7.20	0.06865	3.15	0.64	2.010
4	RC318	210.40	229.80	177.20	52.60	19.40	0.118	0.01300	10.44	0.08279	3.12	0.64	5.226
5	R114	238.07	270.80	172.27	98.53	32.73	0.750	0.08900	8.42	0.14575	3.01	0.61	4.910
6	R11	290.34	331.30	163.06	163.44	40.66	0.408	0.04030	10.12	0.40500	4.02	0.82	8.220
7	R123	216.20	254.70	106.20	148.50	38.50	0.379	0.03300	11.53	0.44443	3.86	0.79	13.600
8	R717	1681.00	2005.00	764.90	1240.50	324.40	3.377	0.43790	7.07	0.28970	3.82	0.78	0.770
9	R718^b	2500.00	3329.00	292.90	3036.00	829.00	0.032	0.00063	51.14	206.30000	3.66	0.75	294.000

$^aV_{rel}$ = relative swept volume of a (piston) compressor, defined as a ratio of swept volume of a compressor working on an arbitrary fluid to the swept volume of a compressor for the R22 working fluid, assuming that revolutions of both compressors are the same; in the case of a heat pump delivering power $Q_{hs} = 10$ kW at 1016 rev/min, the swept volume of a compressor working on R22 is 178 cm³.

^{b}This working fluid is not used in heat pumps at the relevant temperature range; it is given here for comparison only.

Table 4.2   *Suitabilities of different working fluids for the implementation of non-ideal reverse Rankine cycle within temperature range 273–343 K*

Criterion of suitability	Degree of suitability		
	Good (4)	*Quite good* (3.5)	*Adequate* (3)
Large value of the COP	R11 R717 R123	R12 R22	R134c RC318 R114
Advantageous pressure range	R12 R134a	R22 R717	R114 R11 R123 RC318
Low compression ratio	R22 R12 R717 R134a	R114	R11 RC318
Small compressor	R22	R12	R114 RC318 R717 R11 R123

Order of suitability to realize the cycle

No.	Fluid	Average mark
1	R22, R134a	3.8
2	R12, R17	3.7
3	R11, R143	3.3
4	R114	3.2
5	RC318	3.0

Order of suitability to realize the cycle taking into account the ODF

No.	Fluid	Average mark
1	R124a	3.8
2	R717	3.7
3	R22	3.6
4	R123 RC318	3.2
5	R114	1.0
6	R12, R11	0

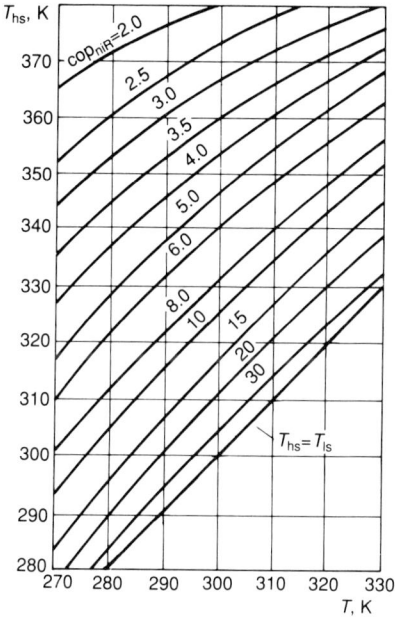

Figure 4.6 *Results of calculations for a non-ideal reverse Rankine cycle employing the R12 working fluid*

An analytical formula for the COP of a non-ideal reverse Rankine cycle is given in [42] as follows:

$$\frac{COP_{niR}}{COP_{niR} - 1} = \frac{T_{hs}}{T_{ls}} \left\{ \frac{1 - \ln{(T_{hs}/T_{ls})}(T_{hs}c_{wL}/r)}{1 - [(T_{hs}/T_{ls}) - 1](T_{hs}c_{wL}/r)} \right\} \tag{4.11}$$

This relationship contains two physical parameters of the substance used: specific heat of the liquid (it is approximately independent of pressure and temperature) and heat of evaporation of the working fluid at the high-temperature heat source temperature level. Figure 4.8 gives the values of specific heat of the liquid at the boiling line (Figure 4.8a) and of the heat of evaporation (Figure 4.8b).

Ziegler [43] presents the following expression for the heat collected from the evaporator:

$$q_{ls} = i_1 - c_{wL}(T_{hs} - T_{ls}) = r - c_{wL}(T_{hs} - T_{ls}) \tag{4.12}$$

It is easy to prove this relationship, since the area $a_3a_235a_4a_3$ (Figure 4.4a) is equal to $c_{wL}(T_{hs} - T_{ls})$, and part of this area, namely area a_235a_2 (Figure 4.4) is equal to the area $a_14a_2a_3a_1$, reducing by this amount the heat collected from the evaporator. An advantage of Equation 4.12 is the possibility of determining the location of point 4 (Figure 4.4a) without the need to follow the throttling process.

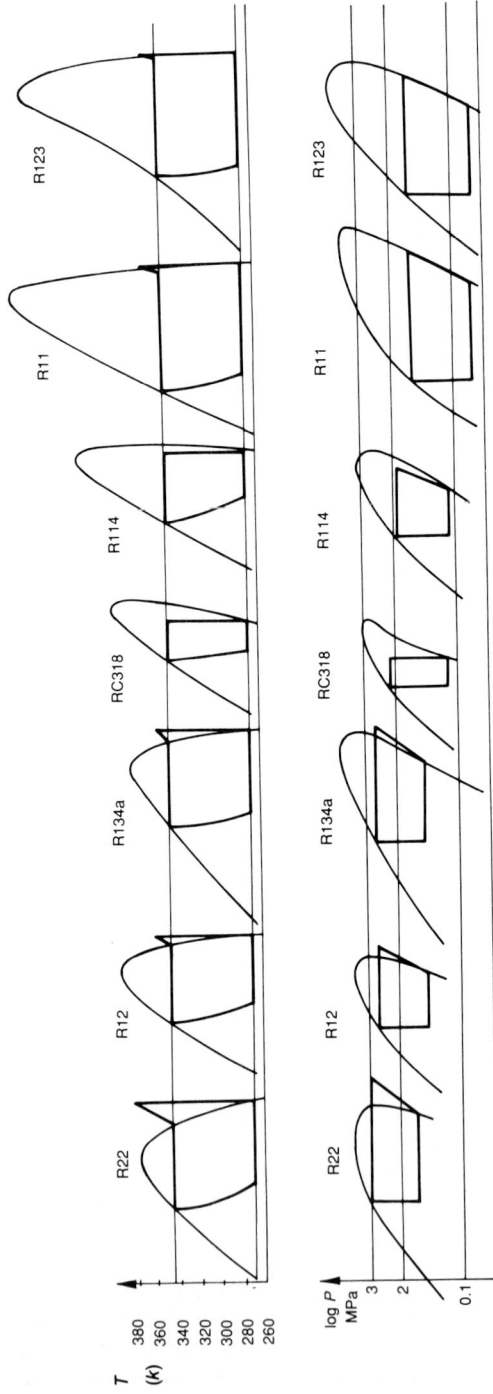

Figure 4.7 The non-ideal reverse Rankine cycles shown in T–s and log P–i coordinates, employing different working fluids within the same temperature range 273–343 K

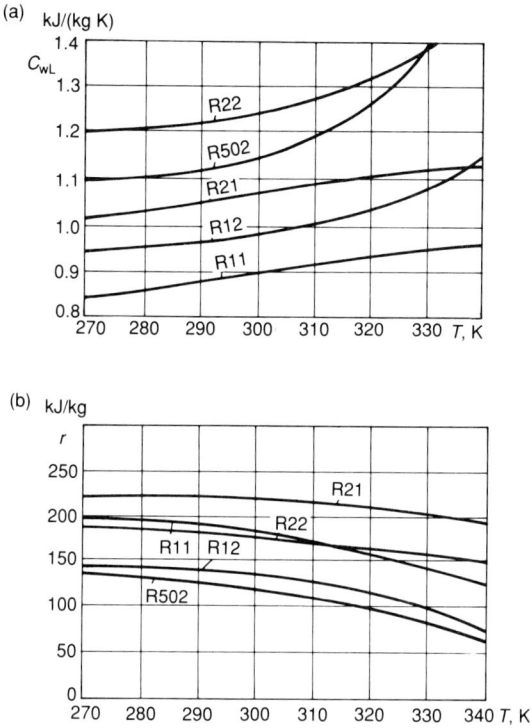

Figure 4.8 *Material parameters of the working fluids, for use in Equation 4.11;* (a) *specific heat of liquid on the boiling line;* (b) *heat of vaporization*

The same point of the cycle (point 4 in Figure 4.4a) can be determined from a different expression. And so, taking into account that area $a_2 35 a_2$ (Figure 4.4a) is equal to the area $a_1 4 a_2 a_3 a_1$ (Figure 4.4a), the following relationship applies:

$$(s_{a_2} - s_5) \frac{T_{hs} - T_{ls}}{2} = (s_4 - s_{a_2}) T_{hs} \tag{4.13}$$

Hence:

$$s_4 = s_{a_2} + \frac{1}{2} (s_{a_2} - s_5)(1 - T_{hs}/T_{ls}) \tag{4.14}$$

Equations 4.10, 4.11 and 4.14 are based on a simplifying assumption that the specific heat does not depend on temperature or pressure. This is true for ideal gases; for liquids the errors resulting from these assumptions are usually smaller than 2%.

4.3 The approximate cycle

4.3.1 Methods of describing individual losses

Development of a method of describing each loss requires a detailed analysis of its parent irreversible process. This in turn requires specifying the equipment in which each process is implemented, isolating the sources of losses, and then making assumptions which make the quantitative description of losses possible. These assumptions can be neither accurate nor unambiguous. However, the very isolation and description of losses create conditions for their minimization. We concentrate here on introducing the methods of description, which can be made more precise and verified as required.

A description of the losses in irreversible processes basically resolves itself into an estimation of the irreversible increase of entropy; having estimated the latter it is not difficult to estimate the exergy loss or the work lost. Here we analyse the losses of a typically thermodynamic character, related to the processes of heat transfer, flows at small pressure differences and throttling (flows at large pressure differences). Apart from these, there are also losses occurring during transmission of work and electrical losses in the motor. Although in the last two cases there are specific losses of the irreversible processes of friction or the flow of electrical current, we will not involve ourselves in a detailed description of these, but will use a ready-made description in the form of efficiencies: mechanical and electrical.

Losses occurring in the condenser and evaporator can be discussed together because of the identical nature of their irreversibilities, induced by the heat transfer at finite temperature differences and by the flow of fluids at finite pressure differences (Figure 4.9).

The description of exergy losses resulting from the heat transfer at finite temperature differences is very simple if these temperatures are constant. The exergy loss is defined by the following expression:

$$\delta B = q \left(\frac{1}{T_1} - \frac{1}{T_2} \right) T_{en} \tag{4.15}$$

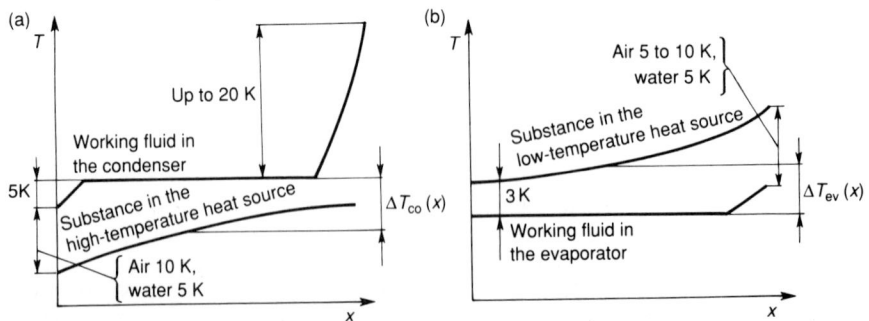

Figure 4.9 *Temperature distributions along the length of condenser and evaporator:* (a) *condenser;* (b) *evaporator*

If the temperatures of substances in the sources are variable, as is the case in a heat exchanger, the losses might be approximately estimated by introducing a temperature defined by the following relationship:

$$T_m = \left(\frac{\Delta i}{\Delta s} \right)_{P = \text{const.}} \tag{4.16}$$

which is analogous to a thermodynamic temperature defined for the working fluid.

Exergy losses resulting from the resistance to flow are defined by the following relationship:

$$\delta b = - T_{en} \int \frac{v}{T} \, dP \tag{4.17}$$

For an isenthalpic throttling process, which takes place in the expansion valve, $T \, ds = v \, dP$ and these losses can be described by the following expression:

$$\delta b = - T_{en} \int ds \tag{4.18}$$

The losses occurring in a compressor depend on its design, and the analysis presented here refers to a piston compressor. Let us concentrate on determining the losses which on the one hand are of typically thermodynamic character, and on the other hand are significant because of their magnitude. Here one can include losses in the compressor valves (which can be simplified to losses in an expansion valve), and losses of heat transfer between the working fluid and the surfaces of cylinder walls and piston. The largest is the heat transfer loss. Mechanical and electrical losses, on the other hand, are treated in a much simplified way.

Two kinds of losses occur in the compressor valves. The first loss is associated with throttling of gas during its flow through the valve, and the second is caused by the work required to open the valves. The description of a throttling loss in the compressor valve is similar to that of a loss in a throttling valve (discussed in Section 4.2). Losses in both the inlet and outlet valve are governed by the following relationship:

$$\Delta s = \delta s \tag{4.19}$$

i.e. all the entropy changes ΔS are due to irreversible losses δS. The entropy changes themselves can be estimated from isenthalpic processes occurring within the assumed range of pressure drop. For valves used in piston compressors, pressure drops can be estimated from the following equation, relating the pressure drop to the total pressure:

$$\delta P_{hs} = 0.1 P_{hs} \tag{4.20}$$

The losses of work in the expansion, inlet and outlet valves are illustrated in Figure 4.10.

Additionally, work related to the necessity of overcoming the spring force and opening the valve is converted into heat. One can thus identify a loss here, although its description would have to be done within the domain of mechanical design procedures. Such a description would have to take into account the characteristics and stiffness of the spring, design of the valve, etc. These problems are not under

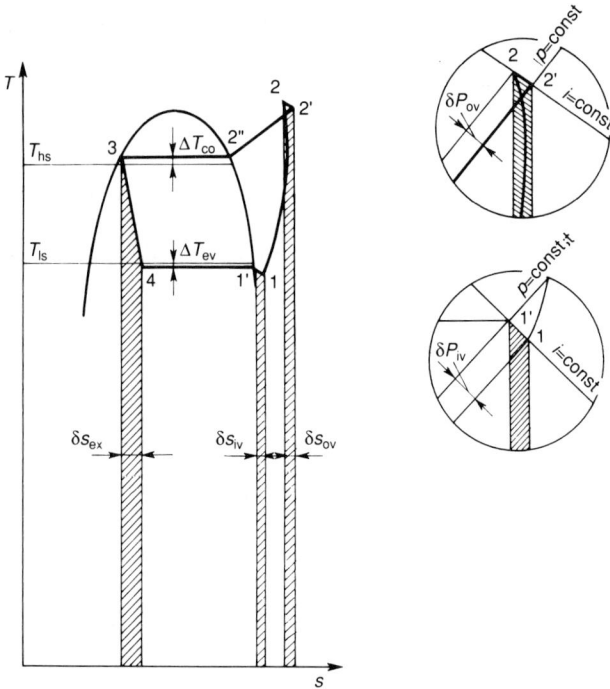

Figure 4.10   *Losses in the expansion valve and in the compressor inlet and outlet valves*

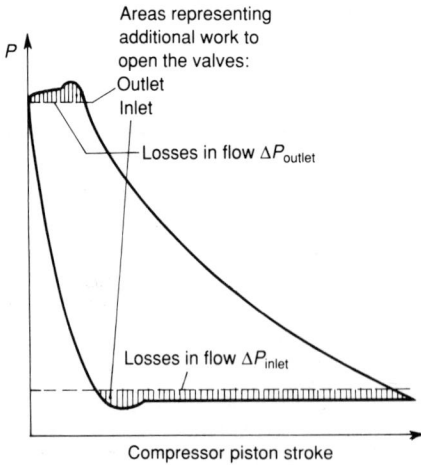

Figure 4.11   *Losses in flow of the working fluid through compressor valves*

Table 4.3 *Characteristic parameters of an exemplary cycle of a mechanical vapour compression heat pump*

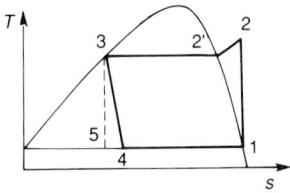

Ideal reverse Rankine cycle 122'3541

Non-ideal reverse Rankine cycle 122'341

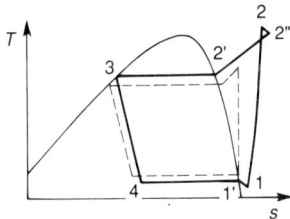

Approximate cycle 1'122"2'34

Salient points of the non-ideal reverse Rankine cycle

Points	i (kJ/kg)	s (kJ/(kg K))	P (MPa)	T (K)
1	654.73	1.5563	0.363	278.0
2	686.51	1.5563	1.707	345.7
2'	676.15	1.5377	1.707	338.0
3	565.42	1.2103	1.707	338.0
4	565.42	1.2315	0.363	278.0

Salient points of the approximate cycle

Points	i (kJ/kg)	s (kJ/(kg K))	P (MPa)	T (K)
1	652.54	1.5655	0.327	270.0
1'	652.54	1.5585	0.363	273.0
2'	677.24	1.5358	1.706	343.0
2	697.79	1.5882	1.896	367.0
2"	697.79	1.5952	1.706	362.0
3	571.21	1.2268	1.706	343.0
4	571.21	1.2607	0.363	273.0

Parameters of the compressor: bore 100 mm, stroke 100 mm, six cylinders; swept volume 4.7 $\times 10^{-3}$ m³; revolutions $n = 591$ rev/min.

consideration here; the loss itself is only mentioned and illustrated on an indicator diagram (Figure 4.11). One can add that the loss is not large, as it occurs only during short periods of time, and thus it can be assumed that its effect is reflected in the mechanical efficiency of a compressor.

Losses related to heat exchange between the working fluid and cylinder walls in the compressor are more difficult to describe quantitatively. The nature of such losses is similar to that of losses during heat transfer in heat exchangers, but here the processes are transient and take place in a complex geometry which additionally varies in time due to the motion of the piston. Vapour entering the cylinder early in the induction stroke is cooler than the walls, so that it absorbs heat, which leads to an increase of the specific volume and temperature of the vapour. Towards the end of induction the temperature of vapour in the cylinder is higher than that of fresh vapour entering, so that the mass flow of the induced vapour is smaller than that which would result from the temperature of fresh vapour upstream of the compressor. As these phenomena are cyclically repeated, in order to estimate the exergy losses one needs to integrate in time and space both the energy equation for the working fluid and the unsteady state heat conduction equation for the cylinder wall. Without going into details of these relatively complex calculations, the resulting magnitudes of losses, for a compressor described in Table 4.3, are presented in Table 4.4.

Table 4.4   *Calculation of entropy increases*

Increase due to:	Formula	Value (kJ/(kg K))	Necessary assumptions
Heat transfer			
in the condenser	$\delta s_{co} = q_{hs}(T_m - T_{hs})/(T_m T_{hs})$	0.0060	$\delta T_{co} = 5$ K
in the evaporator	$\delta s_{ev} = q_{ls}(T_{ls} - T_{ev})/(T_{ls} T_{ev})$	0.0053	$\delta T_{ev} = 5$ K
Throttling in the valve			
expansion	$\delta s_{ex} = \Delta s = s_4 - s_3$	0.339	
compressor outlet	$\delta s_{ov} = \Delta s = s_{2'} - s_2$	0.0070	$\delta P_{ov} = 0.1 P_{hs}$
compressor inlet	$\delta s_{iv} = \Delta s = s_1 - s_{1''}$	0.0070	$\delta P_{iv} = 0.1 P_{ls}$
Heat transfer in the compressor cylinder	$\delta s_{wc} = s_2 - s_1$	0.0227	
Entropy transferred through the cylinder walls	$s_{wc} = q_w/T_w$	0.0004	$T_w = 340$ K area of fins 0.1 m², coefficient of heat transfer from fins to air 25 W/m² K)

Friction losses and losses of work needed to drive auxiliary equipment, such as the valves of an oil pump, which are dissipated into thermal energy, can be partially or wholly transferred to the high-temperature heat source. These losses can thus be described by the entropy increase. The issue is not, however, developed here any further, and these losses are all included in the mechanical efficiency (η_m). Also present in the drive system are electrical losses. These are not transferred as heat to the high-temperature heat source (usually the motor is installed outside the space being heated), but result in an increased requirement for the driving energy. These losses can also be described by the electrical efficiency (η_{el}).

4.3.2 The method of adding losses (a numerical example)

In the previous section we have presented methods of describing the individual losses. The description of losses, followed by their superposition, enables the formulation of an approximate cycle, which is the key task. The aim is to have at our disposal a description which is as close as possible to the cycle practically implemented in a heat pump.

Table 4.5 *Exergy losses and driving exergy losses in the approximate cycle*

Losses in:	$m \, \delta b$ (kW)	$m \, \delta b / W_a$ (%)
Condenser	1.5429	3.67
Evaporator	1.3629	3.24
Expansion valve	8.7174	20.76
Compressor outlet valve	1.8001	4.29
Compressor inlet valve	1.8001	4.29
Compressor cylinder	5.8373	13.90
External loss (exergy transferred to the environment)	0.0248	0.06
Subtotal	21.0855 = 21.09	50.21
Exergy transferred at the high-temperature heat source	20.9000	49.79
Total	41.9900	100.00

The 'superposition' of losses has to take into account various conditions important in the method. First of all, when estimating individual losses, one has to take into account that some of them have an effect on some others. Thus the process of superposition begins at the stage of estimating the individual losses, which have thus to be calculated in a particular order of succession. Secondly, the comparative coefficients must refer to cycles *having the same effects* (Section 4.1). Thus these coefficients need to be somewhat rescaled. The condition of this rescaling can be expressed as follows. If the exergy supplied in the high-temperature heat source of an approximate cycle is the same as that of, for example, an ideal reverse Rankine cycle, then the following condition must be complied with:

$$Q_{iRhs} = Q_{ahs} \tag{4.21}$$

For an approximate cycle in which the variable temperature of the working fluid along the length of the condenser is replaced by a temperature defined by Equation 4.16, the above condition leads to:

$$Q_{ahs} = Q_{niRhs} = Q_{Chs} \tag{4.22}$$

In this correlation the Carnot cycle is defined by a high-temperature heat source temperature equal to the temperature T_m.

Table 4.6 *Comparison of equivalent cycles*

Cycle	Q_{hs} (kJ)	W (kJ)	M (kg)	δB (kJ)	COP
Carnot	117.09	20.90	1.000	0.00	5.60
Ideal reverse Rankine	117.09	20.90	1.000	0.00	5.60
Non-ideal reverse Rankine	117.09	28.46	1.000	7.56	4.11
Approximate	117.09	41.99	0.925	21.09	2.79
Practically realized	117.09	47.02	0.925	26.12	2.49

Since in an approximate cycle exergy is lost in the condenser, the cycle has to be 'rescaled' by the quantity of mass of the working fluid, according to the equation:

$$M(\Delta b_{ahs} - \delta b_{co}) = b_{iRhs} \tag{4.23}$$

where Δb_{ahs} denotes the change of exergy of the fluid discharging the heat at the high-temperature heat source.

Heat discharged at the high-temperature heat source is equal to the sum of heat absorbed at the low-temperature heat source, exergy losses dissipated as heat and work supplied to the heat pump:

$$Q_{ahs} = Q_{als} + w_{iR} + M\sum_j \delta b_j \tag{4.24}$$

The work of an approximate cycle is equal to:

$$W_a = w_{ir} + M\sum_j \delta b_j \tag{4.25}$$

The work of a practically realized cycle, assuming that its corresponding approximate cycle describes all losses in the heat pump installation faultlessly, exceeds the work of such an approximate cycle by the losses in the compressor and in the electric motor. This work can be calculated from the following formula:

$$W_r = W_a/(\eta_m \eta_{el}) \qquad (4.26)$$

Comparing the approximate and practically realized cycles with other cycles, assuming that $T_{ls} = T_{en}$, yield the following correlations for the comparative coefficients:

$$\varphi_{a,niR} = W_{niR}/W_a$$

$$\varphi_{a,iR} = W_{iR}/W_a = \eta_{ac}^* \qquad (4.27)$$

$$\varphi_{r,iR} = W_{iR}/W_r = \eta_{rc}^*$$

A full presentation of the approximate cycle and obtaining an estimate of its losses is only possible using a numerical example. The presentation comprises: a formulation of the description of the cycle, estimation of losses and comparative coefficients. The following conditions were assumed: temperature of the high-temperature heat source for the non-ideal reverse Rankine and ideal cycles was estimated according to Equation 4.16 as 338.41 K. The temperature of the low-temperature heat source was assumed to be $T_{ls} = T_{en} = 278$ K. The values of mechanical and electrical efficiency are, respectively, $\eta_m = 0.94$ and $\eta_{el} = 0.95$. The calculations were performed for the R–12 working fluid. Numerical values obtained from the solution of this example are given in Tables 4.3 to 4.7, and the Saukey diagrams for energy and exergy are shown in Figure 4.12.

Table 4.7   *Cycle comparative coefficients*

Coefficient	Definition	Value
$\varphi_{a,niR}$	W_{niR}/W_a	0.678
$\varphi_{a,iR} = \varphi_{a,C} = \eta_{ac}^*$	W_C/W_a	0.498
$\varphi_{niR,iR} = \varphi_{niR,C} = \eta_{niR}^*$	W_C/W_{niR}	0.734
$\eta_{ac}^* = \varphi_{niR,iR}\,\varphi_{a,niR}$	W_C/W_a	0.498
$COP_{ac} = \eta_{ac}^*\,COP_C$	Q_{hs}/W_a	2.790
$\varphi_{r,a}$	W_a/W_r	0.893
$COP_{rc} = \varphi_{r,a}\,COP_{ac}$	Q_{hs}/W_r	2.490
$\eta_{rc}^* = \varphi_{niR,iR}\,\varphi_{a,niR}\,\varphi_{r,a}$	W_C/W_r	0.444

The numerical values assumed in this example and the results obtained can be supplemented by the following comments, which may be useful in developing a better understanding of the calculation procedure and results.

Comment 1. The basic task is to estimate the parameters of the Carnot, non-ideal reverse Rankine and ideal reverse Rankine cycles. Losses, on the other hand, can be calculated in parallel with the calculation of parameters of the approximate cycle, following a certain order, which is not maintained at the stage of presenting the results. The description of individual losses is also an open issue.

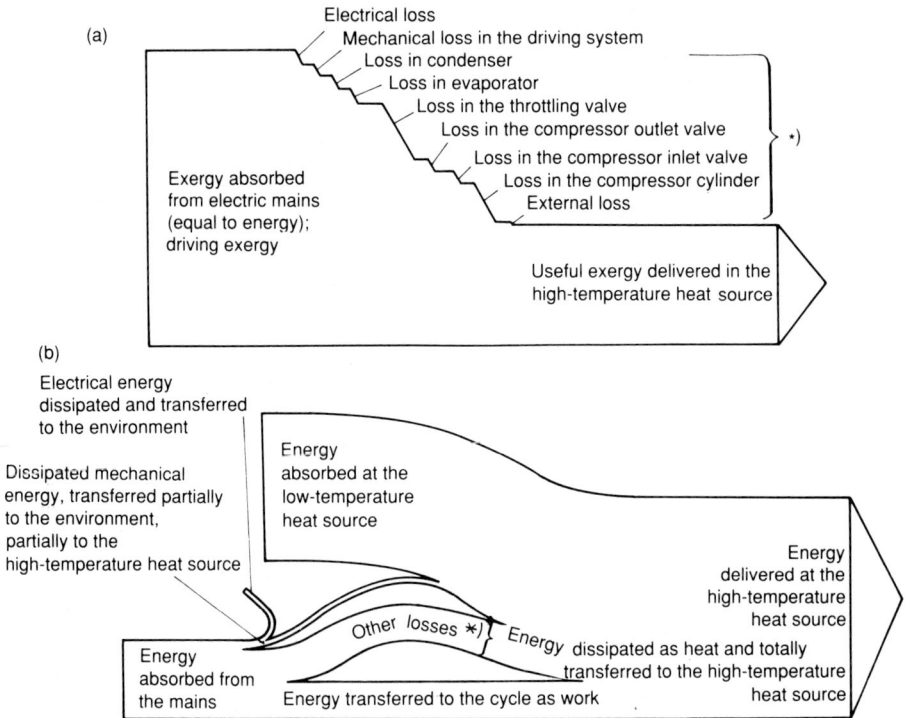

(a)

Electrical loss
Mechanical loss in the driving system
Loss in condenser
Loss in evaporator
Loss in the throttling valve
Loss in the compressor outlet valve
Loss in the compressor inlet valve
Loss in the compressor cylinder
External loss

*)

Exergy absorbed
from electric mains
(equal to energy);
driving exergy

Useful exergy delivered in the
high-temperature heat source

(b)

Electrical energy
dissipated and transferred
to the environment

Energy
absorbed at the
low-temperature
heat source

Dissipated mechanical
energy, transferred partially
to the environment,
partially to the
high-temperature heat source

Energy
delivered at the
high-temperature
heat source

Other losses *)

Energy dissipated as heat and totally
transferred to the high-temperature
heat source

Energy
absorbed from
the mains

Energy transferred to the cycle as work

Figure 4.12   *Flow diagrams for:* (a) *exergy;* (b) *energy*

Comment 2. The proposed temperature difference is somewhat extreme. Differences larger than 40 or 50 K are used with reluctance. Such a large difference (60 K) was proposed deliberately to exaggerate the losses; hence relatively large losses were obtained.

Comment 3. The methods of describing losses were considered to be more important than the numerical results, although efforts were made to use typical values. Thus the accuracy of the presented parameters results from the accuracy of the calculations, but does not reflect the precision of the method used. Only the Carnot cycle can be calculated with unlimited accuracy. The calculations of the non-ideal reverse Rankine cycle cannot be too accurate, as different tables and charts give differing data on the properties of working fluids. Obtaining high accuracies when calculating the losses is difficult.

4.4. Thermodynamic and economic optimization of a mechanical compression heat pump

The exergetic losses, discussed in Section 4.3.2, should be minimized for thermodynamic reasons, so that the driving exergy can be transformed into useful exergy with maximum efficiency. This leads to lower costs of the driving exergy. Alas, the minimization of losses requires the use of better equipment, and thus additional capital costs. A problem defined in such a way resolves into an issue of the combined thermodynamic and economic optimization. Heat pumps differing in power, type of compressor used, working fluid employed, kind of driving energy used, etc. could be the subject of such analysis. However, before any quantitative description and thus economic analysis is attempted, a number of decisions narrowing the task have to be taken. Further on in the book an analysis, based on [44], is presented of a small mechanical compression heat pump, designed to work in a family house as a source of hot water and space heating. The power in the high-temperature heat source is 6.5 kW, temperatures $T_{hs} = 333$ K (60°C) and $T_{ls} = 290$ K (17°C), the working fluid is the R12 refrigerant, the pump is driven by an electric motor, and the estimated life of the pump is nine years. The task was to design a pump which would be as cheap as possible to build and run, using the elements available on the Swedish market (compressors, motors, condensers and evaporators), selected to provide the desired economic effect. This is a typical problem for the practice of optimization, which means that both the method of formulation of a problem and methods of solution are known. The way to proceed is to formulate a target function, defining the costs of individual equipment and installation, depending on the exergetic losses within this equipment and on the costs of driving energy. The extreme value of this function is found next, taking into account the known limitations of the decisive variables. Due to their character, the solution is based on the method of Lagrange multipliers. The multipliers can also have some economic significance, which allows us to analyse the effect of prices of individual elements of the installation on its total cost.

The formulation of a function defining the costs of individual devices depending on exergetic losses occurring within these devices is a key task. One needs to know the market prices of standard pieces of equipment such as condensers, evaporators, electric motors and compressors. It is relatively easy to develop such a cost function for electric motors, as the motor efficiency (and it is an exergetic efficiency) is a standard specification quoted by the manufacturers. It is more difficult to define such a function for compressors of various designs. The definition of efficiency can be limited to the dominant loss within the cylinder. The condenser and evaporator

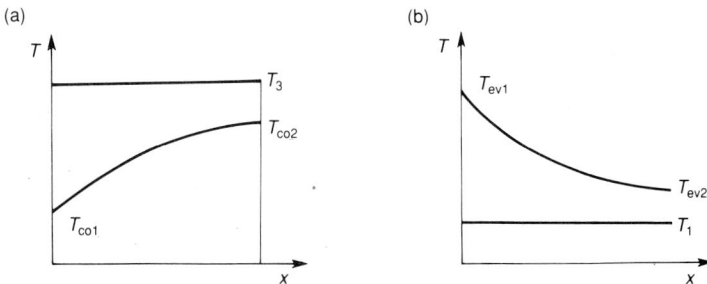

Figure 4.13 *Simplified temperature distribution in:* (a) *the condenser;* (b) *the evaporator*

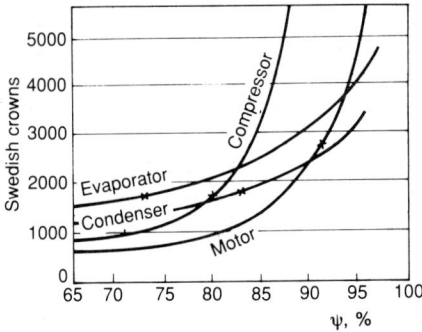

Figure 4.14   *Changes in capital costs (column* (c) *in Table 4.8) as a function of the value of a decisive function which is a measure of perfection of the equipment; points on curves mark the optimum values*

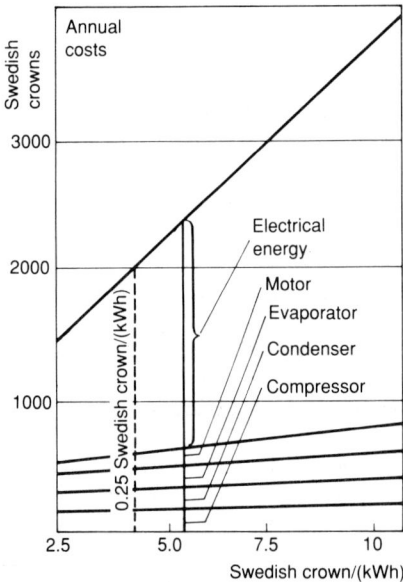

Figure 4.15   *Optimize annual costs of operating the heat pump (assuming its depreciation spread over nine years) as a function of the costs of electrical energy*

are in fact heat exchangers, in which exergy losses can be estimated accurately. The authors of the reported work have, however, used a simplified description. They have made use of the fact that exergy losses are directly proportional to the temperature difference at which heat transfer takes place. A ratio of temperature differences of the condensing vapour and of the cooling fluid (Figure 4.13) provides a measure of exergy losses for the condenser. A similar ratio is used for the evaporator. Such a measure of losses can also be used as a measure of the exchanger costs, since a zero value corresponds to an exchanger of zero surface area, and the value of one to an infinitely large exchanger; the cost of the exchanger being directly related to its surface area.

The decisive variables and the form of a cost function adopted in the reported work [41] are listed in Table 4.8. The limitations, in the form of equations, are: heat flux at the high-temperature heat source $Q'_{hs} = 6.5\,kW$, at temperature $T_{hs} = 333\,K$, assumed number of working hours 5000 per annum, an electrical energy cost, according to the then actual costs in Sweden, equal to 0.25 Swedish crowns per 1 kWh. Table 4.8 also lists some results of the optimization calculations, and Figure 4.14 shows the changes of capital costs corresponding to the changes in decisive variables, including their optimum values (marked by points on the curves). Figure 4.15, on the other hand, shows the running costs of the heat pump, for different prices of electrical energy, with depreciation spread over nine years.

For the calculated optimum conditions: electric motor power is 1.796 kW, the losses total $0.954\,kW$, $COP_{rc} = 3.78$, exergetic efficiency $\eta^*_{rc} = (1.796 - 0.954)/1.796 = 0.468$.

4.5 Types of motors used to drive the compressors; regeneration of heat losses* using a heat engine

Compressors are usually driven by electric motors, and work losses which could be recovered in the form of heat are small at the heat pump location. When heat engines are used, much larger possibilities exist for recovering (regenerating) the cooling and exhaust losses. This does not increase the COP of the heat pump cycle as such, but increases the COP of the whole installation (including the engine). The problem of choosing the engine type has several aspects: technical, thermodynamic, but first of all economic.

The technical aspects refer to the possibility of using the drive which is optimum in the given conditions. The problems associated with such a choice are only mentioned here. The engine or motor has to be available on the market, it needs access to an energy delivery system, suitable space to be installed in, availability of spare parts, servicing, oils, greases, etc. The effects of the engine on the environment (emission of exhaust gases or noise) may also be important. Other important aspects are dimensions of the heat pump and of the engine driving it, and the ease of operating and controlling the engine.

In the case of a small heat pump, an electric motor might easily comply with these demands, but for larger installations there might be limitations, e.g. on the electrical mains power available. In such cases the use of an internal combustion engine may be beneficial. A preferred choice is to use a standard traction engine running on diesel fuel, which is both proven in terms of its reliability and relatively cheap. Where a dual tariff is used for electrical energy, a dual drive might be

---

*This is a commonly used name; strictly speaking this is energy transferred to the environment: energy of flue gases in the form of internal energy, energy of cooling in the form of heat.

Table 4.8 Assumptions and results of the optimization calculations

Device	Assumptions		Results of optimization calculations		Energy losses		
	Decisive variables	Cost functions	Optimum value of decisive variable	Cost of device (Swedish crowns)	W	% lost of electric power	% of total loss
Compressor	$\psi_1 = \dfrac{\Delta i_{is}}{\Delta i_r}$	$c_1 = a_1 k_1 v_1 \dfrac{1}{(0.9 - \psi_1)} \left(\dfrac{P_2}{P_1}\right) \ln\left(\dfrac{P_2}{P_1}\right)$	$\psi_1 = 0.80$	1650	268	15	28
Condenser	$\psi_2 = \dfrac{T_{co2} - T_{co1}}{T_3 - T_{co1}}$	$c_2 = a_2 k_2 m_2 \sqrt{\left(\dfrac{\psi_2}{1 - \psi_2}\right)}$	$\psi_2 = 0.83$	1787	123	7	13
Expansion valve	–	–	–	–	295	16	31
Evaporator	$\psi_3 = \dfrac{T_{ev1} - T_{ev2}}{T_{ev1} - T_1}$	$c_3 = a_3 k_3 m_3 \sqrt{\left(\dfrac{\psi_3}{1 - \psi_3}\right)}$	$\psi_3 = 0.73$	1677	76	4	8
Electric motor	$\psi_4 = \dfrac{N_m}{N_{el}}$	$c_4 = a_4 k_4 P \sqrt{\left(\dfrac{\psi_4}{1 - \psi_4}\right)}$	$\psi_4 = 0.91$	2585	192	1	20
Total					954	53	100

Key: a_i, technical aging indicator including the costs of cleaning, repairs, etc.; k_i, costs of flow; m_i, costs of flow; v_1, specific volume of vapour.

profitable: electric motor during the cheap electricity tariff period and internal combustion engine at other times. Large mechanical compression heat pumps require the use of steam or gas turbines to drive the compressor. The manufacture of such turbines is quite complex and costly.

Quite a number of references deal with thermodynamic aspects related to various forms of energy and their conversion to mechanical energy at the compressor shaft: [45], [46], [47], [2], [48], [49], [30]. The data given are neither too general nor too precise. Here we will limit ourselves to presenting two proposed solutions (the numerical data quoted should be treated with some reserve).

The first solution includes a discussion whether it is better to use an electric motor or an internal combustion engine to drive the heat pump. According to [45], [2] and [50], the internal combustion engine is better, because of the primary energy of fuel used in such engines (100%); assuming engine efficiency of 0.34, regeneration of the cooling and exhaust losses at 0.75, and COP of the heat pump of 3.5, one can obtain 170% of the primary energy at the high-temperature heat source level. On the other hand, when an electric motor is used, at the assumed power station efficiency 0.38, the efficiency of transmission of electrical energy 0.9 and the heat pump COP is again equal to 3.5, this quantity is only 120%. These numbers result from the fact that using a heat engine to drive the heat pump gives better opportunities of utilization of the primary energy at the heat pump location. The quoted difference in numbers between 170% and 120% results from assuming that losses can be recovered only in the heat engine, disregarding the possibilities of their recovery in the power station (e.g. in district heating).

The second solution is an analysis of driving the heat pump by a heat engine with and without regeneration of heat losses in the engine. The simplifying conditions for this analysis are: both the heat engine and heat pump cycles are described by Carnot cycles with their corresponding exergetic efficiencies, the engine operates within a temperature range T_{ext}, T_{hs}, with an exergetic efficiency η^*_{HE}, and the heat pump operates within T_{hs} and T_{ls}, with an exergetic efficiency η^*_{HP}.

PER of the heat pump, referring to the primary energy, and taking into account the degree of regeneration (σ) of the heat engine waste heat (Q_{hsII}) transferred to the user at the high-temperature heat source level together with heat originating from the heat pump itself, is equal to:

$$\text{PER} = \frac{Q_{hsI} + \sigma Q_{hsII}}{Q_{ext}} \tag{4.28}$$

Introduction of the exergetic efficiencies of the heat pump η^*_{HP} and that of the heat engine η^*_{HE}, leads to:

$$\text{PER} = \eta^*_{HP}\eta^*_{HE} \frac{T_{hs}}{(T_{hs} - T_{ls})} \frac{(T_{ext} - T_{hs})}{T_{ext}} \left[1 + \sigma \frac{T_{hs} - T_{ls}}{T_{hs}} \frac{1}{\eta^*_{HP}} \right.$$
$$\left. \left(\frac{T_{ext}}{T_{ext} - T_{hs}} \frac{1}{\eta^*_{HE}} - 1 \right) \right] \tag{4.29}$$

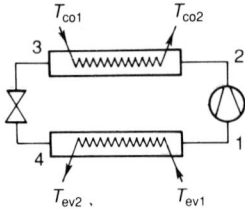

Results of calculations for $T_{ext} = 1500\,\text{K}$, $T_{ls} = 273\,\text{K}$, $\eta^*_{HP} = 0.6$, $\eta^*_{HE} = 0.4$ are presented in Figure 4.16. The presented analysis is to a large extent based on assumptions of quantitative character. The whole problem here lies rather with the proper choice of coefficients of various kinds than with the thermodynamic calculations. The unit cost of electrical energy must be such that payback is within three to four years. In most developing countries electrical energy is expensive and in such cases the application of heat pumps is questionable.

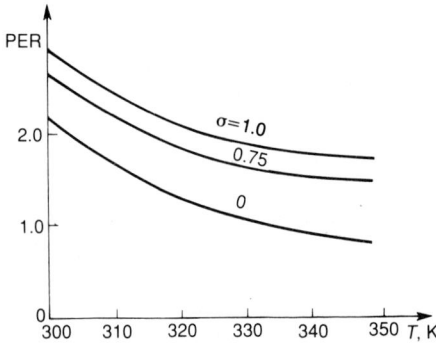

Figure 4.16 *Primary energy ratio for different values of the degree of regeneration*

A technical implementation of regeneration of heat in the installation of a mechanical compression heat pump, driven by a steam turbine employing heat regeneration in the turbine condenser, is presented in Example 6.8.

In the extreme case of $\eta^*_{HP} = \eta^*_{HE} = 1$ and $\sigma = 1$, all heat from the low-temperature heat source of the heat engine is transferred to the high-temperature heat source of the heat pump. Such conditions might be assumed for a sorption cycle heat pump. Equation 4.29 reduces then to the correlation (see also Equation 5.3):

$$\text{PER} = \text{COP}_{ia} = \frac{T_{hs}}{T_{hs} - T_{ls}} \frac{T_{ext} - t_{ls}}{T_{ext}} \tag{4.30}$$

This correlation is also valid for the ideal cycle of an ejector heat pump discussed in detail in Example 6.3.

Chapter 5

Sorption cycle heat pumps

5.1 Description of a sorption heat pump cycle

In Chapter 1 we explained the principles of operation of the sorption cycle heat pumps: absorption cycle heat pump, heat transformer and a resorption cycle heat pump driven by mechanical or thermal energy. Resorption heat pumps utilize a cycle similar either to the mechanical vapour compression heat pump (Figure 1.4) or to the absorption heat pump (Figure 1.6), so the description of an absorption heat pump or a heat transformer can be considered as the leading one within the range of sorption-type heat pumps. A review of the literature on sorption cycle heat pumps is given in [51].

Suitable comparative cycles are introduced in the quantitative description of a sorption cycle heat pump. A first approximation is provided by an ideal cycle (Figure 5.1), consisting of a dual Carnot cycle or a dual Ericson cycle. For the theoretical cycle providing the second approximation, a cycle is proposed which depends on the kind of working fluid sorbent employed and takes into account the exergy losses resulting from irreversible heat transfer in the regenerative exchangers, the rectification column and throttling in the valves. An extension of this cycle, called the approximate cycle, additionally takes into account exergy

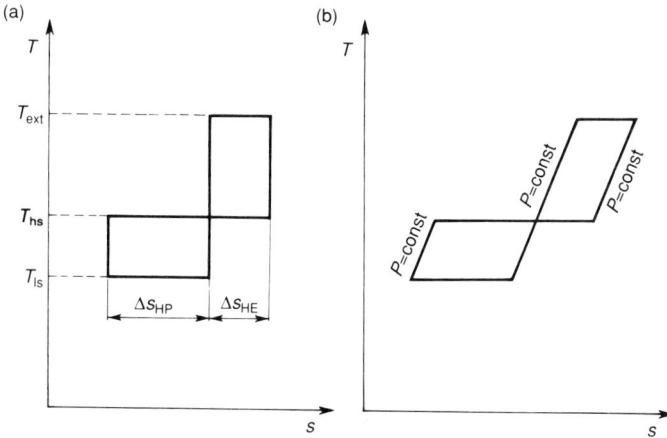

Figure 5.1 *Ideal cycles of an absorption cycle heat pump:* (a) *dual Carnot cycle;* (b) *dual Ericson cycle*

losses in heat exchangers at the heat sources and exergy losses resulting from flow resistances.

The COP of an ideal cycle, according to Equation 1.3 and adopting the nomenclature of Figure 5.1, can be defined as follows:

$$COP_{ia} = \frac{T_{hs} (\Delta s_{HP} + \Delta s_{HE})}{T_{ext} \Delta s_{HP}} = \frac{T_{hs}}{T_{ext}} \frac{\Delta s_{HP}}{\Delta s_{HE}} + \frac{T_{hs}}{T_{ext}} \tag{5.1}$$

The second term in this equation results from the fact that the bottom temperature of the heat engine is equal to the top temperature of the heat pump. Taking into account that the engine work is transferred to the actual heat pump cycle,

$$(T_{ext} - T_{hs}) \Delta s_{HE} = (T_{hs} - T_{ls}) \Delta s_{HP} \tag{5.2}$$

and hence:

$$COP_{ia} = \frac{T_{hs}}{T_{hs} - T_{ls}} \frac{T_{ext} - T_{hs}}{T_{ext}} + \frac{T_{hs}}{T_{ext}} = COP_C \, \eta_{HE} + \frac{T_{hs}}{T_{ext}} \tag{5.3}$$

This correlation is often given in the following form (Figure 5.2):

$$COP_{ia} = \frac{T_{hs}}{T_{hs} - T_{ls}} \frac{T_{ext} - T_{ls}}{T_{ext}} \tag{5.4}$$

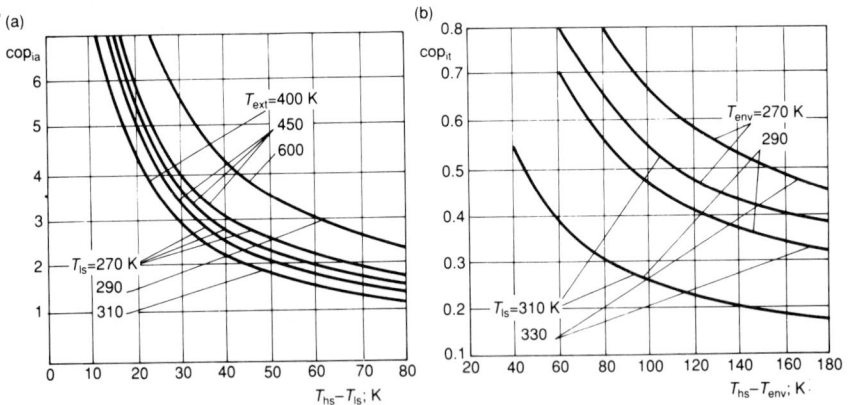

Figure 5.2   *Coefficient of performance of an ideal cycle of a sorption heat pump as a function of temperature differences:* (a) *absorption cycle heat pump;* (b) *heat transformer*

The COP for a heat transformer may be written similarly:

$$COP_{it} = \frac{T_{hs}}{T_{hs} - T_{env}} \frac{T_{ls} - T_{env}}{T_{ls}} \tag{5.5}$$

The theoretical and approximate cycles depend on the kind of working fluid and sorbent employed. It is customary to illustrate these cycles either on the $i = i(z)$ chart at $P = $ const. and $T = $ const., or on the log P–1/T diagram at $z = $ const. The absorption cycles of refrigeration installations are discussed in [52], and those of heat pumps in [53] and [54].

Both the intensive and extensive parameters of the cycle, including the heat delivery and reception, can be represented on an $i = i(z)$ chart. It should be stressed, however, that the cycle of a sorption heat pump cannot be presented in a

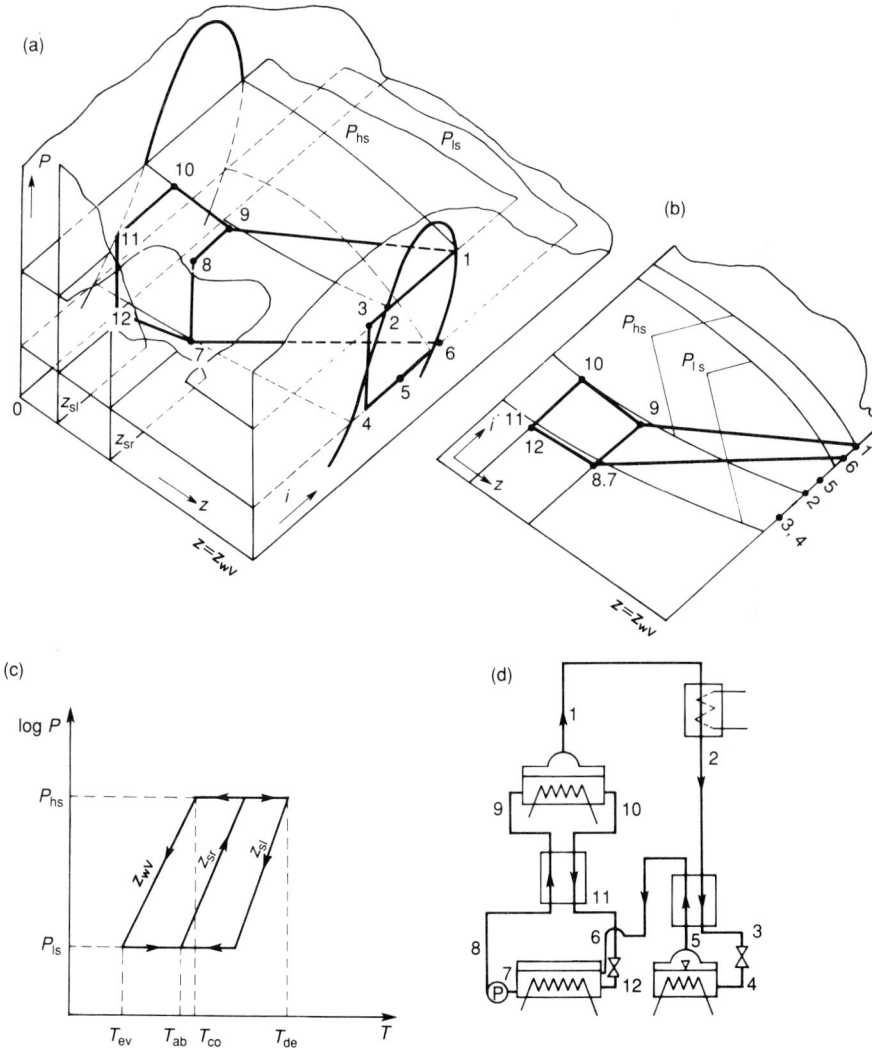

Figure 5.3 *The absorption heat pump cycle:* (a) *on the* $i = i(P, z)$ *chart;* (b) *in the* i–z *plane;* (c) *in the* log P–1/T *plane;* (d) *diagram of the installation*

Table 5.1 *Graphical principles of balances*

No.	Fluxes being added

W = resultant flux.

Balance in the i–z plane

Continuity equation: $A + B - W = 0$
Mass balance equation for the component: $Az_A + Bz_B - Wz_W = 0$
Energy equation: $Ai_A + Bi_B - Wi_W = 0$

Variant I: $A > B; A - B - W = 0$
Variant II: $A < B; A - B + W = 0$

Continuity equations, other equations as in row 1

$A + B - C - D = 0$
Variant I: $A + B = C + D; A + B = W_1; C + D = W_1$
Variant II: $A - C = D - B; A - C = W_2; D - B = W_2$
Variant III: $C - B = A - D; C - B = W_3; A - D = W_3$

Continuity equations, other equations as in row 1

$Ai_A + Bi_B + Q - Wi_W = 0$
Variant I: $A(i_A + Q/A) + Bi_B = Wi_W$
Variant II: $Ai_A + B(i_B + Q/B) = Wi_W$
Variant III: $Ai_A + Bi_B = W(i_W - Q/W)$

Energy equation, other equations as in row 1

$Ai_A + Bi_B - Q/A - Wi_W = 0$
Variant I: $A(i_A - Q/A) + Bi_B = Wi_W$
Variant II: $Ai_A + B(i_B - Q/B) = Wi_W$
Variant III: $Ai_A + Bi_B = W(i_W + Q/W)$

---

form commonly used in the theory of thermodynamic cycles (according to which a thermodynamic cycle consists of consecutive changes of a constant quantity of a working fluid). This results from the fact that the $i = i(z)$ chart is a simplified form of a three-dimensional $i = i(z, P)$ chart, presented (together with the cycle itself) in Figure 5.3a. On a two-dimensional chart (Figure 5.3b) some processes applied to the working fluid partly coincide; for example, 1 and 2 coincide with 4 and 5 (condensation and evaporation), or degenerate to a single point, e.g. 3 and 4, 7 and 8, 11 and 12 (isenthalpic throttling of the fluid, pumping of the solution, isenthalpic

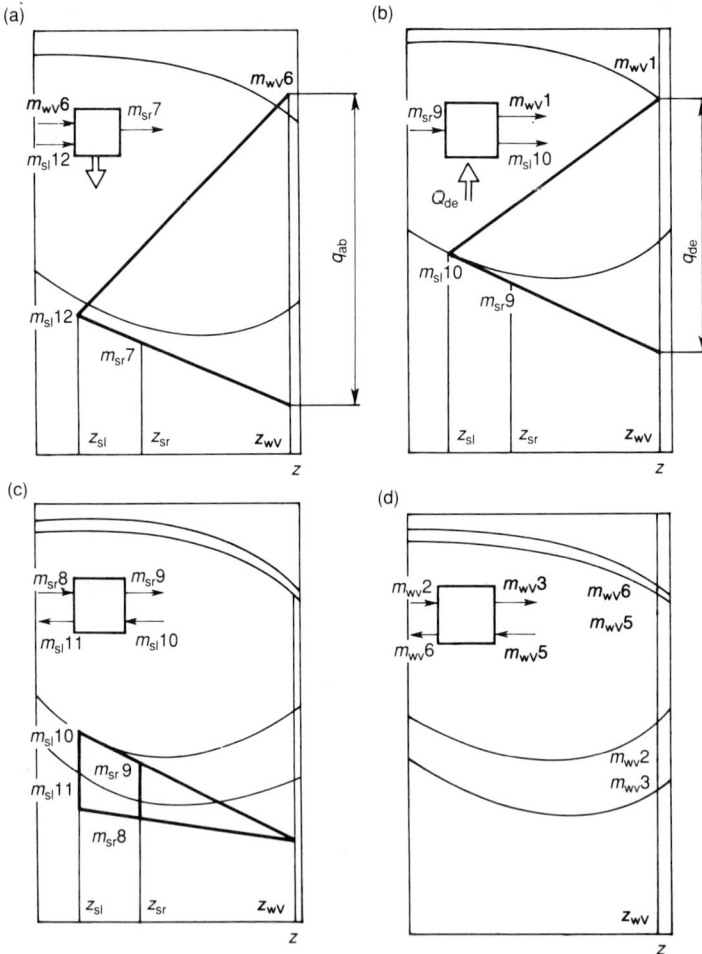

Figure 5.4 *Balances of the individual nodes of a heat pump installation, executed following the graphical balancing rules introduced in Table 5.1 (numbers denominating the individual thermodynamic states correspond to* **Figure 5.3d**):
(a) absorber; (b) desorber; (c) regenerative heat exchanger for the solutions;
(d) regenerative heat exchanger for the vapours

throttling of the rich solution). This makes difficult the interpretation of the chart and presentation of the cycle.

Two cycles can be identified within a sorption heat pump cycle: the actual heat pump cycle and the heat engine cycle (each of these involves different quantities of fluids). The actual heat pump cycle 1, 2, 3, 4, 5, 6 is an open cycle, consisting of a constant pressure condensation on subcooling of the condensate, isenthalpic throttling, constant-pressure evaporation and superheating of the working fluid vapour. The heat engine cycle is also an open cycle 6, 7, 8, 9, 1, consisting of the constant-pressure vapour absorption, pumping of the rich solution and constant-pressure desorption. The lean solution circulates between the desorber and absorber (points 10, 11, 12). Salient points of the cycle define only the intensive parameters; for a full description a knowledge of the mass flows of the solution is required. These values, referred to 1 kg of the working fluid, can be obtained from the mass balance equations. The chart $i = i(z)$ makes such balancing possible, as both its coordinates are pseudointensive parameters.

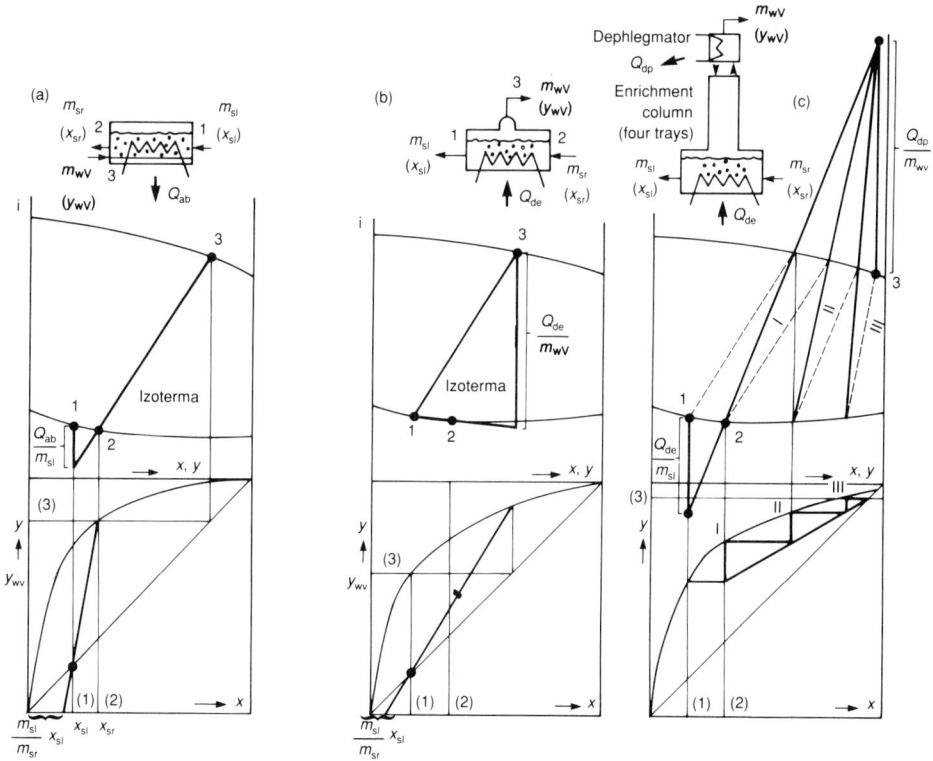

Figure 5.5 *Thermodynamic states of the solution and working fluid in the absorber and desorber:* (a) *tank (mixed) absorber, liquid leaving the tank is in equilibrium with the feed vapour;* (b) *tank (mixed) desorber (commonly used), vapour leaving the desorber is in equilibrium with the liquid;* (c) *desorber with a column and dephlegmator giving better separation of the working fluid and sorbent*

The thermodynamic cycle of a sorption heat pump can also be presented in the log P–$1/T$ plane, which makes possible a determination of intensive parameters of the cycle. In such a plane the magnitudes of heat fluxes are proportional to the slopes of lines of constant z (Section 2.2).

The rules of mass balance in the i–z plane (without accumulation) are simple. Summation of the two flows of the substance ($A + B$, Table 5.1) leads to the three equations of conservation; of the total mass ($A + B = C$), of the mass of working fluid ($Az_A + Bz_A = Cz_A$) and of exergy ($Ai_A + Bi_A = Ci_A$). The balances at some more important points of the installation are shown in Figure 5.4. Figure 5.5 shows an interpretation of the mass transfer processes in the absorber and desorber depending on their design.

For these pairs of working fluid–sorbent, which display relatively large differences of vapour pressures of the working fluid and sorbent (z_{wV} close to unity), the effect of desorber design is not as large as might be suggested by Figure 5.5.

5.2. Theoretical cycle of a sorption heat pump

The cycle of a sorption heat pump employs a two-phase, two-component working fluid (Chapter 1). It is assumed that its fraction in the vapour leaving the desorber is equal to unity. This means that condensation and evaporation are realized using a single-component fluid. Assuming further that the thermodynamic states of solutions leaving the desorber and absorber are located on the corresponding lines representing the boiling liquid, one can calculate the number of degrees of freedom characterizing the cycle. According to the Gibbs rule, applied four times, this number equals six. Additionally, the following equalities are valid in the installation: $P_{de} = P_{co}$ and $P_{ab} = P_{ev}$ (neglecting the resistance to flows); thus for the given temperatures of the sources T_{ls} and T_{hs}, two degrees of freedom remain, and some liberty is allowed concerning their choice. For an absorption heat pump

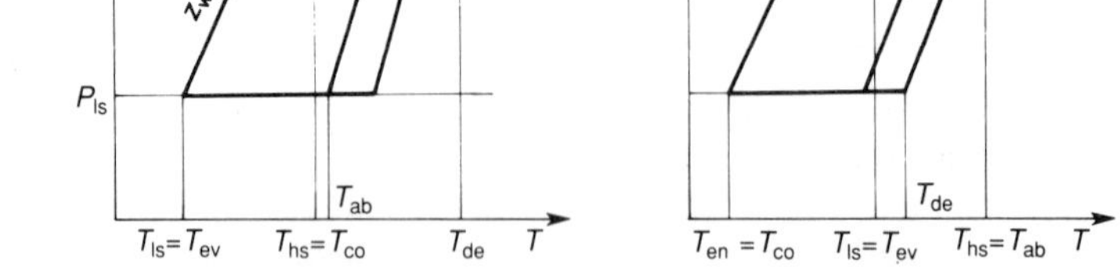

Figure 5.6   *Theoretical cycles of a sorption cycle heat pump:* (a) *absorption cycle heat pump;* (b) *heat transformer*

one assumes an absorber temperature T_{ab} and a difference in mass fractions Δz; for a heat transformer the desorber temperature and difference in mass fractions are assumed.

Figure 5.6 illustrates the design of cycles for an absorption heat pump and a heat transformer. For the absorption heat pump, the assumption of temperatures $T_{ls} = T_{ev}$ and $T_{hs} = T_{co}$ determines the pressures $P_{ev} = P_{ls}$ and $P_{co} = P_{hs}$. The intersection of a constant-pressure line P_{ls} with the isotherm T_{ab} determines the value of the mass fraction $z_{sr} = z_{sl}(T_{ab}, P_{ls})$, and the value of mass fraction $z_{sl} = z_{sr} - \Delta z$ determines the desorber temperature $T_{de} = T_{de}(z_{sl}, P_{hs})$.

The quantities of heat exchanged (per 1 kg of working fluid vapour) can be determined graphically (as in Figure 5.4) or by solving a set of conservation equations. For the absorption cycle heat pump this set has the following form:

Mass balance equation for the total
    mass: $m_{wV} + m_{sl} = m_{sr}$

Mass balance equation for the
    working fluid: $m_{wV}z_{wV} + m_{sl}z_{sl} = m_{sr}z_{sr}$

Energy conservation equations: $q_{ab} = (r+1)_{ab} + q_{se} + q_{ve} + (f-1)V_{sl}\,\Delta P$

Absorber: $q_{de} = (r+1)_{de} + q_{se} + c_{wL}(T_{de} - T_{ab}) -$
                                         q_{dp}

Desorber: $q_{ev} = r_{ev} - c_{wV}(T_{de} - T_{ab}) + q'_{ve} - v_{wL}\,\Delta P$

Evaporator: $q_{co} - r_{co} + c_{wV}(T_{de} - T_{co})$

Solution pump: $w_{sr} = fv_{sr}\,\Delta P$ (5.6a)

For a heat transformer the mass balance equations for total mass and for the working fluid are the same as for an absorption heat pump, and the energy conservation equations are as follows:

Absorber: $q_{ab} = (r+1)_{ab} - q_{se} - c_{wV}(T_{ab} - T_{ev})$

Desorber: $q_{de} = (r+1)_{de} - q_{se} - c_{wL}(T_{ab} - T_{de}) - q_{dp} - fV_{sr}\,V\Delta P$

Evaporator: $q_{co} = r_{co} + c_{wL}(T_{ev} - T_{co})$

Condenser $q_{co} = r_{co} + c_{wV}(T_{de} - T_{co})$

Solution pump: $w_{sl} = (f-1)v_{sl}\,\Delta P$

Working fluid pump: $w_{wL} = v_{wL}\,\Delta P$ (5.6b)

A heat transformer installation usually does not include a regenerative heat transformer for the vapours ($q'_{ve} = 0$). It was assumed in the above equations that the specific heat and specific volume of solutions vary linearly with the fluid fraction, so that the following relationships are valid:

$$c_{wL} = c_{sl}f - c_{sr}(f-1) \tag{5.7a}$$

$$v_{wL} = v_{sl}f - v_{sr}(f-1) \tag{5.7b}$$

where f denotes the ratio of mass flows of the rich solution and working fluid. The first two equations of the set (5.6a) result in a correlation:

$$f = m_{sr}/m_{wV} = (z_{wV} - z_{sl})/(z_{sr} - z_{sl}) \tag{5.8}$$

The individual terms in these equations have different physical meanings. For the absorber and desorber the heat of phase change is correspondingly equal to the heat of evaporation or condensation augmented by the heat of dissolution (mixing). The heat q_{se} is contained in the lean solution, flowing in the solution heat exchanger (for the heat pump). In an infinitely long exchanger all this flux is transferred to heating of the rich solution, but due to the finite length of the exchanger, only a part of this flux is transferred. The non-transferred part of this flux is equal to:

$$q_{se} = (1-\varepsilon_{se})(f-1)c_{sl}(T_{de} - T_{ab}) \tag{5.9}$$

where ε_{se} denotes the efficiency of the heat exchanger. For an infinitely long exchanger, $\varepsilon_{se} = 1$ and $q_{se} = 0$.

The heat q_{ve} is contained in the condensate flowing in the vapour heat exchanger. The part of this heat which is not transferred to the vapours, is:

$$q_{ve} = \varepsilon_{ve}c_{wv}(T_{co} - T_{ev}) \tag{5.10}$$

where ε_{ve} is the efficiency of this exchanger.

The real (non-latent) heat flux in the energy equation for the desorber results from different values of heats and mass flows, and q_{dp} describes the heat consumed in the dephlegmator (condenser) of the rectifier column. Work delivered to the mechanical pump $fv_{sr}\Delta P$ is dissipated in the expansion and equalizer valves.

For example, from the above analysis it follows that satisfying the condition $T_{ab} = T_{co}$ demands that $\varepsilon_{ve} = 1$.

The analysis of effects of the degrees of freedom on the COP can be started by assuming $T_{ab} = T_{co}$ and $\Delta z = 0$. This requires a further assumption that the designs of absorber and desorber will guarantee sufficiently large retention volume of solutions that the sorption of the working fluid vapour will not affect the constant value of its fraction in the solution. Additionally, mass fluxes of solutions have to be infinitely large compared with the fluid vapour flux ($f = \infty$), and the temperature of the lean solution entering the absorber must be equal to the temperature of absorption. This leads to a demand for the solution heat exchanger to be infinitely long, $\varepsilon_{se} = 1$. A similar conclusion can be drawn regarding the demands put on the regenerative heat exchanger for the vapours due to the assumed temperature difference $T_{ab} = T_{co}$. With these assumptions the terms q_{se} and q_{ve} in equations 5.6a become zero. Further analysis of this particular case of a theoretical cycle also neglects the power delivered to the solution pump.

The cycle contains irreversible entropy increases resulting from heat exchange in the regenerative exchangers for vapours and solutions, heat and mass exchange in the rectification column, heat exchange in the desorber and mass exchange in the absorber. The absorption process is conducted at constant pressure and temperature, but at a finite difference of mass fractions. This results in the necessity to use a rectification column to separate vapours of the working fluid from the vapours which are in equilibrium with the boiling lean solution in the desorber. In the rectification column, heat and mass transfer are taking place at a finite difference of temperatures and mass fractions, which results in an irreversible entropy increase. Such irreversible entropy increases cause an increase in the demand for driving exergy, and thus also for thermal energy, compared with the ideal cycle. This increased demand can be characterized by calculating the value of the exergetic efficiency coefficient. Using the definition of the COP

$$(\text{COP})_{ta} = (q_{ab} + q_{co} + q_{dp})/q_{de} \qquad (5.11)$$

with values shown in Figure 5.7 as a function of temperature, the exergetic efficiency coefficient can be derived as

$$\eta^*_{ta} = \text{COP}_{ta}/\text{COP}_C \qquad (5.12)$$

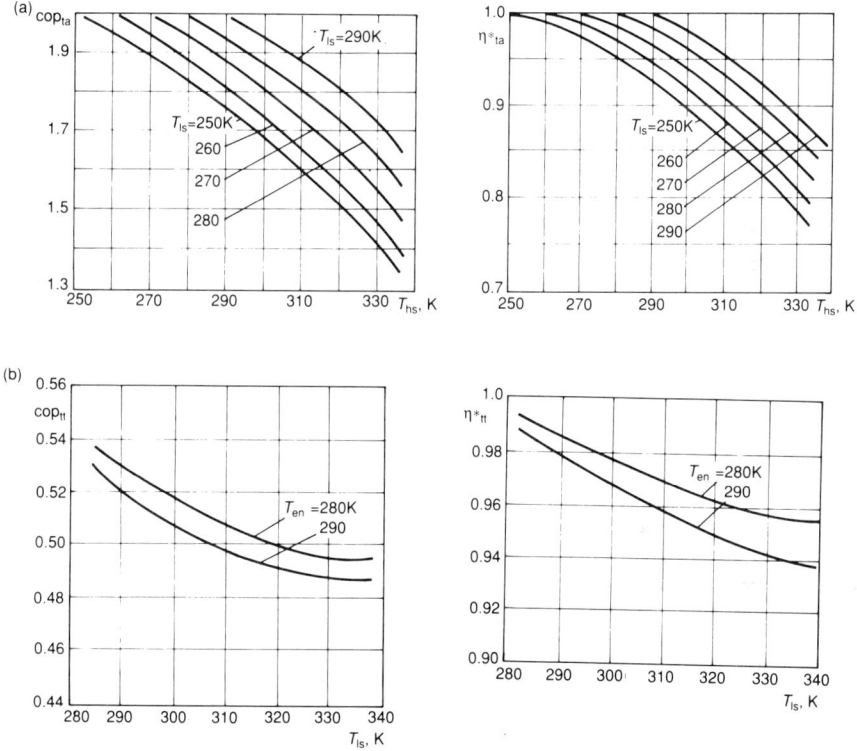

Figure 5.7 *Coefficients of performance and exergetic efficiency:* (a) *absorption cycle heat pump employing the R717 (ammonia)–R718 (water) pair;* (b) *heat transformer employing the R718 (water)–lithium bromide pair*

Temperatures T_{de} (for the sorption heat pump) and T_{co} (for the heat transformer) found for this cycle are correspondingly the lowest and the highest allowed temperatures (Figure 5.8). The cycle discussed above also makes it possible to postulate a condition at which the exergetic efficiency will become unity. As can be seen in Figure 5.8 for the absorption cycle heat pump, when T_{hs} approaches T_{ls}, temperature T_{de} also approaches T_{ls} and the mass fraction of working fluid in the solution approaches unity. This means that the absorber has

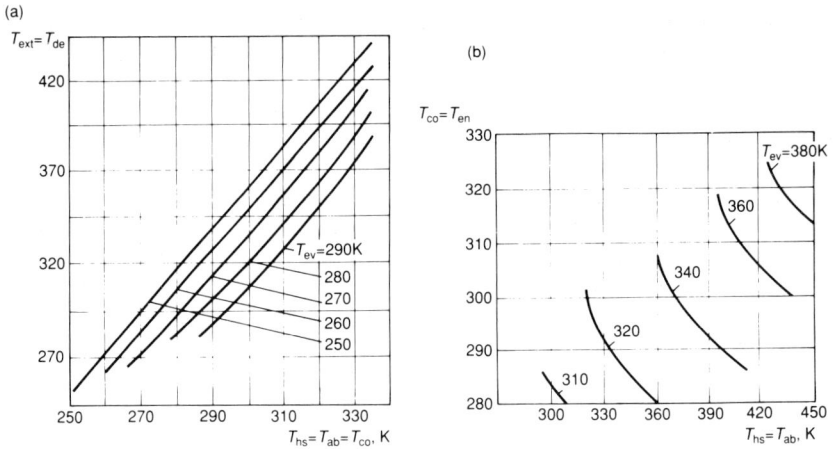

Figure 5.8   *Temperatures of the sources of a sorption heat pump:* (a) *minimum temperature $T_{ext} = T_{de}$ for an absorption cycle heat pump using a R717 (ammonia)–R718 (water) pair;* (b) *maximum temperature $T_{co} = T_{en}$ for a heat transformer with the R718 (water)–lithium bromide pair*

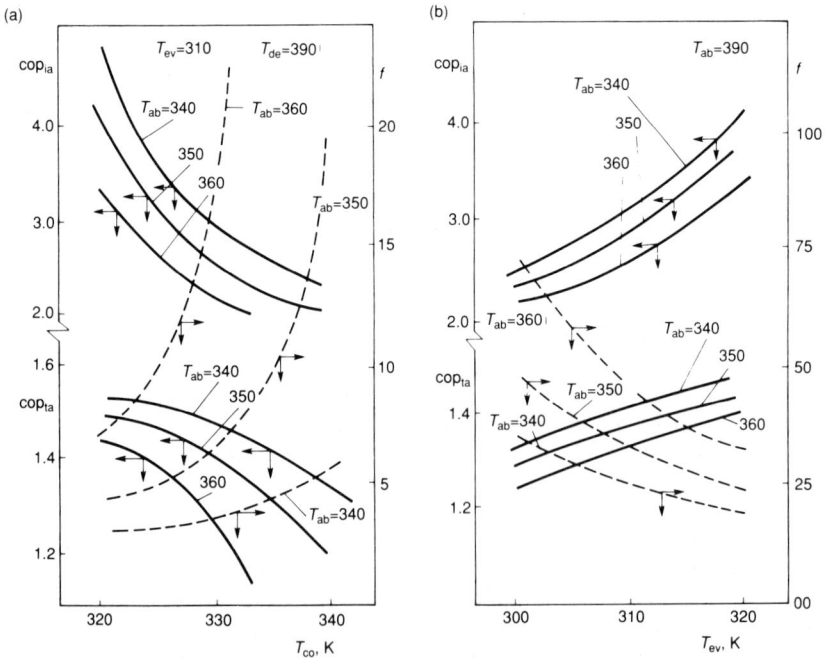

Figure 5.9   *Results of calculations for a theoretical cycle of an absorption cycle heat pump using the R717–water pair [55]:* (a) *COP_{ta} and COP_{ia} as a function of T_{co};* (b) *COP_{ta} and COP_{ia} as a function of T_{ev}*

performed the function of a second condenser, and the desorber that of a second evaporator. For this limiting condition of temperature values there is then no need to use any regenerative heat exchangers or a rectification column. Hence the exergy losses approach zero, and the exergetic efficiency approaches unity. Additionally, an obvious equality is fulfilled: $q_{co} = q_{ab} = q_{de} = q_{ev}$, $q_{dp} = 0$ and the COP reaches the limiting value $COP_{ta} = 2$. The limiting value of the COP for the heat transformer is equal to 0.5. Larger values can be obtained in a multistage installation, which is discussed, among others, in [43]. The assumed degrees of freedom have limiting values, due to the demands on both the thermodynamic parameters and the design of exchangers. A better approximation of the cycle can be obtained by assuming $T_{ab} \neq T_{co}$ and $\Delta z \neq 0$. The results of calculations are shown in Figure 5.9 and in Table 5.2.

A method of increasing the COP by using adiabatic sorption processes is presented in [56] and [57].

Table 5.2   *Exemplary results of calculations of a theoretical cycle of an absorption cycle heat pump: $T_{ls} = 290$ K, pair R717–water*

T_{ab}	T_{co}	T_{de}	z_{sr}	z_{sl}	f	COP_{ta}	η_{ta}^*
330	320	470	0.4820	0.0937	2.33	1.29	0.334
330	330	380	0.4820	0.4638	29.46*	1.76	0.566
330	340	470	0.4820	0.1413	2.52	1.30	0.410
340	330	470	0.4249	0.0937	2.74	1.48	0.425
340	340	400	0.4249	0.4221	206.40*	1.87	0.615
340	340	470	0.4249	0.1413	3.03	1.26	0.484

*Parameters of this cycle are closest to those of a cycle in which $\Delta z = 0$; η_{ta}^* has a maximum value.

5.3 The approximate cycle

The approximate cycle is realized in an installation similar to that of a theoretical cycle. Limited heat capacities of the substance in the sources are assumed, and hence heat exchange between the working fluid and substance of the sources takes place at finite temperature differences (Chapter 3). Additional exergy losses are thus present. Determination of these losses, leading to the calculation of the COP, constitutes an important element of a description of the cycle, similarly to the case of a mechanical vapour compression heat pump (Chapter 4).

A different method of determining the losses is used for the sorption cycle heat pump, for two reasons. The first results from the fact that the nature of the majority of losses is the same and is caused mainly by the finite temperature differences at which the heat transfer processes take place in the exchangers (the extent to which these losses are caused by the pressure losses in flow is much smaller). The second results from the possibility of a quantitative description within the domain of the theory of heat and mass transfer, in the form of certain functions of the cycle parameters and not of the numerical values as was the case in Chapter 4.

The method of calculations is a certain generalization of a method presented among others in [58], [59], [60], [61], [46], [62], [63], [64], [65], [66] and [67]. The formulation of exergy losses in the form of certain functions makes it possible – using optimization calculations – to determine the values of parameters of the approximate cycle and of the COP of a sorption heat pump. The set of parameters for this cycle is much larger than that for an ideal reverse Rankine cycle. The theoretical cycle is described mainly by the lumped parameters, i.e. each exchanger which is an intake of the heat source has only one parameter (temperature) allocated to it, and this parameter is determined by external conditions. Temperature distributions in the heat exchangers of an installation of a sorption cycle heat pump, for the approximate cycle, are presented in Figure 5.10. Temperature distributions in the condenser and evaporator are omitted for simplicity.

The large number of parameters characterizing an approximate cycle is a result of assumptions made with regard to individual processes. In the desorber the process of separating the working fluid (refrigerant) vapours from the solution is limited, so that the leaving working fluid vapour contains also small quantities of the sorbent vapour $z_{wv} < 1$. In the condenser some subcooling of the condensate can occur, and its condition does not correspond to the boiling line. In the

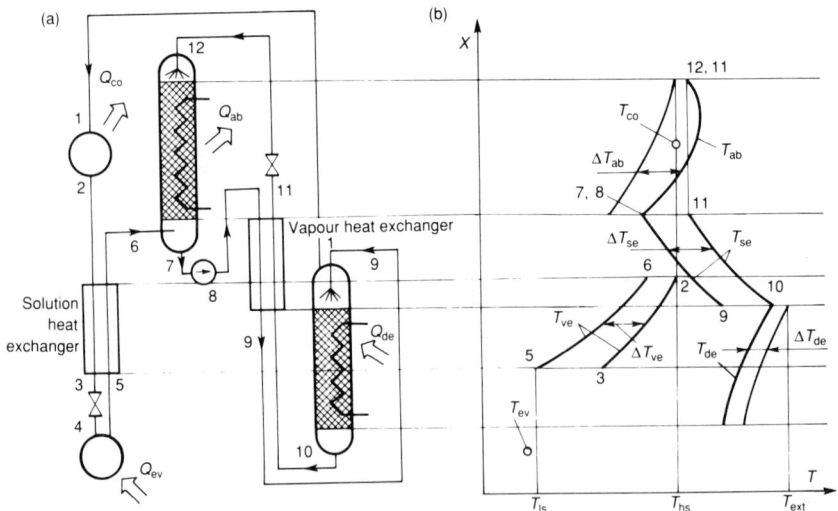

Figure 5.10   *Temperature distributions in the individual heat and mass exchangers of an absorption cycle heat pump installation (as a function of the exchanger length–coordinate z): (a) diagram of the installation (with absorber and desorber providing a continuous contact of the phases, which makes it possible to present temperature distributions along their lengths); (b) temperature distributions in an approximate cycle*

evaporator some entrainment of liquid takes place and the vapour leaving the evaporator is wet. The state of the rich solution does not correspond to the boiling line ($z_{sr} < z_{sr}^*$). Additionally, the state of the solution leaving the desorber, as defined by the constant temperature line (Figure 5.11) does not coincide with the state of the lean solution ($z_{ls} \neq z_{ls}^*$). As the fraction of working fluid in the vapour

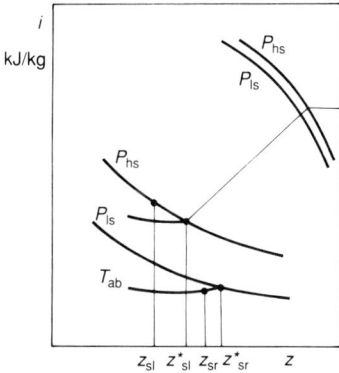

Figure 5.11 *Efficiencies of mass transfer in the absorber and desorber,* $\eta_{ab} = (z_{sr} - z_{sl})/(z_{sr}^* - z_{sl})$, $\eta_{de} = (z_{sl}^* - z_{sl})/(z_{sr} - z_{sl})$

leaving the desorber is smaller than unity, definition of the condensation and evaporation temperatures does not unequivocally define the corresponding pressures, as is the case for a theoretical cycle. The range of variation of the condensation pressure defines the possible subcooling of condensate $P_{co} \varepsilon (P_{co1}, P_{co2})$, where P_{co1} is a pressure corresponding to the condensation temperature of the working fluid characterized by $z = z_{wV} < 1$ (at this pressure the condensate is not being subcooled), and P_{co2} is a pressure corresponding to the condensation temperature of the fluid characterized by $z_{wV} = 1$ (at this pressure the subcooling of condensate is largest). Evaporation pressures are within the range $P_{ev} \varepsilon (P_{ev1}, P_{ev2})$, where P_{ev1} is a pressure corresponding to the absorption temperature T_{ab} and $z = z_{sl}$, and P_{ev2} is a pressure corresponding to to the evaporation temperature T_{ab} at the fraction $z = 1$. The effect of pressures on the value of the COP can be illustrated as follows. Lowering the pressure in the evaporator increases the enthalpy of working fluid, and thus for a fixed value of the heat flux causes a decrease of the fluid mass flux. This decrease, on the other hand, leads to an increase of the pressure difference in the installation and as a consequence may lead to an increase of the power delivered to the solution pump and an increase of the throttling loss. Thus there are optimum values of pressures, for which the losses will have minimum value.

To calculate exergy losses in the approximate cycle, values of the intensive and extensive parameters were established at the individual nodes of the installation.

In a theoretical cycle, external parameters and properties of the working fluid–sorbent pair were sufficient to unequivocally define its thermodynamic states. The approximate cycle is a different case. This cycle is realized in exchangers of finite lengths, and all the heat and mass transfer processes are realized at finite differences of temperature or mass fraction; in addition the flows of fluid and solution are accompanied by pressure losses. Thus to define the relationships between internal and external parameters one needs to know the coefficients characterizing the kinetics of heat and mass transfer, resistances to flow, efficiencies of exchangers etc.

A set of parameters $\mathscr{A}_a$ which characterize the approximate cycle can be presented as follows:

$\mathscr{A}_{a1}$	T_i, P_i, i_i
Internal parameters	m_{sr}, m_{sl}, m_{wV}
$\mathscr{A}_{a2}$	$U_{ab}A_{ab}, U_{co}A_{co}, U_{de}A_{de}, U_{ev}A_{ev}$
Parameters characterizing the kinetics	$U_{se}A_{se}, U_{ve}A_{ve}$
of heat and mass transfer	
$\mathscr{A}_{a3}$	$\Delta T_{ab}, \Delta T_{co}, \Delta T_{de}, \Delta T_{ev}, \Delta T_{se}, \Delta T_{ve}$
Parameters characterizing the exergy	$\eta_{ab}, \eta_{de}, \eta_{el}, \eta_{m,sl}, \eta_{m,sr}, \eta_{m,wl}, \delta T_{ab},$
losses	$\delta T_{co}, \delta T_{de}, \delta T_{dp}, \delta T_{ev}, \delta P_{hs}, \delta P_{ls}$
$\mathscr{A}_{a4}$	$C_{ex}, C_{hs}, C_{ls}, T_{ext}, T_{hs}, T_{ls}$
External parameters	m_{ext}, m_{hs}, m_{ls}

In order to establish the values of internal parameters of the cycle as a function of the external parameters, one needs to make certain assumptions characterizing this cycle. The external parameters are dictated to the heat pump by its function, this defines the temperature and power of the high-temperature heat source and the temperature and power of the low-temperature heat source. Also necessary is an arbitrary assumption of certain temperature differences and coefficients characterizing the efficiency of mass transfer.

The dependency of internal parameters on the external ones can be obtained from the solution of conservation equations for the individual nodes of the installation. For example, in the case of the desorber, using the notation as in Figure 5.12, these equations have the following form:

Equation of conservation of the	
total mass:	$m_6 + m_{10} = m_7 + m_{wv}$
Equation of conservation of the	$m_6 z_6 + m_{10} z_{10} = m_7 z_7 + m_{wv} z_{wv}$
working fluid mass:	
Equation of energy	$Q'_{de} + i_6 m_6 + i_{10} m_{10} = m_7 i_7 + m_{11} i_{11}$ (5.13)
conservation:	

where:

$$Q_{de} = U_{de} A_{de} (\Delta T_{ln})_{de}$$

Exergy losses can be calculated from the equation:

$$\delta B'_{de} = B'_{de} + m_6 b_6 + m_{10} b_{10} - m_7 b_7 - m_{wv} b_{wv} \tag{5.14}$$

where:

$$B'_{de} = Q'_{de}\left(1 - \frac{T_{ext}}{T_{de}}\right) \quad \text{and} \quad b_i = i_i - T_{ext} S_i \tag{5.15}$$

for $i = 6, 7, 10$.

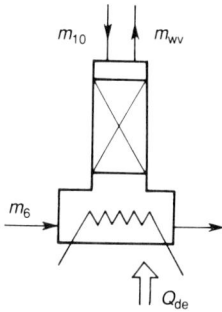

Figure 5.12 *Diagram of a desorber with a rectification column*

Equations of the above type, written for all the nodes of the installation, form a set of non-linear algebraic equations. This set, together with the correlations describing the working fluid and sorbent thermodynamic properties, makes it possible to analyse the effect of individual parameters on the COP of the heat pump. A solution of this set can be attempted using generally available procedures or a method suggested by Dykhevizen [67].

The resulting solution of such a set of equations makes it possible to analyse the effects of individual parameters from sets $\mathcal{A}_{a2}$ and $\mathcal{A}_{a3}$ on the magnitude of exergy losses. Initial values of these parameters are assumed arbitrarily, and then by varying these values (within the ranges resulting from the conditions in which individual processes are implemented), their effect on the internal parameters and

Table 5.3 *Values of parameters in the subset $\mathcal{A}_{a3}$ for an absorption cycle heat pump*

Cause of loss	Parameters
Heat transfer	Temperature drops of the source substances $\delta T_{ev} = 7\text{ K}, \delta T_{de} = 0$
	Increase of temperature of the source substances $\delta T_{ab} = 54.4\text{ K}, \delta T_{co} = 37.7$ $\delta T_{dp} = 14.8\text{ K}$
	Minimum temperature differences at the exchanger inlets or outlets $\Delta T_{ev} = 3.92\text{ K}, \Delta T_{de} = 8\text{ K}$ $\Delta T_{de} = 4.36\text{ K}, \Delta T_{co} = 18.75\text{ K}$ $\Delta T_{dp} = 10.19\text{ K}$
Mass transfer	$\eta_{ab} = \eta_{de} = 0.9$
Resistance to flow	$\delta P_{ls} = \dfrac{P_{ev} - P_{ab}}{P_{ev}} = 0.02$
	$\delta P_{hs} = \dfrac{P_{de} - P_{co}}{P_{de}} = 0.01$

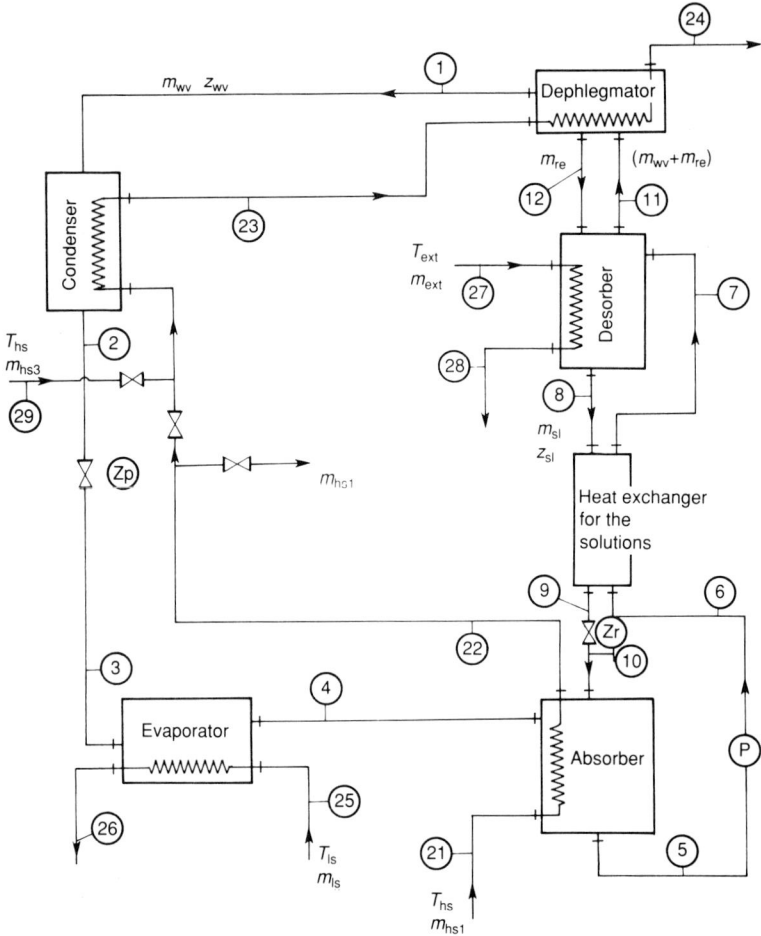

Figure 5.13   *Diagram of the installation of an absorption cycle heat pump: z_p, expansion valve, z_r, equalizing valve*

on exergy losses can be analysed. The optimum choice of these parameters can be obtained through optimization calculations, which demands a formulation of the *target function*. As in any technical project, this function should be determined by the economic effects. The optimization is limited here to establishing only the maximum values of the COP and exergetic efficiency.

The method of calculation for an approximate cycle is illustrated on examples of a heat pump and a heat transformer. Calculations were performed for the pair R717–H$_2$O.

In the absorption cycle heat pump installation, shown in Figure 5.13, water provides the low temperature heat source. Parameters characterizing the exergy losses (subset $\mathcal{A}_{a3}$) are listed in Table 5.3. The calculated internal and external parameters are given in Table 5.4. Figures 5.14a and 5.14b show the flow diagrams

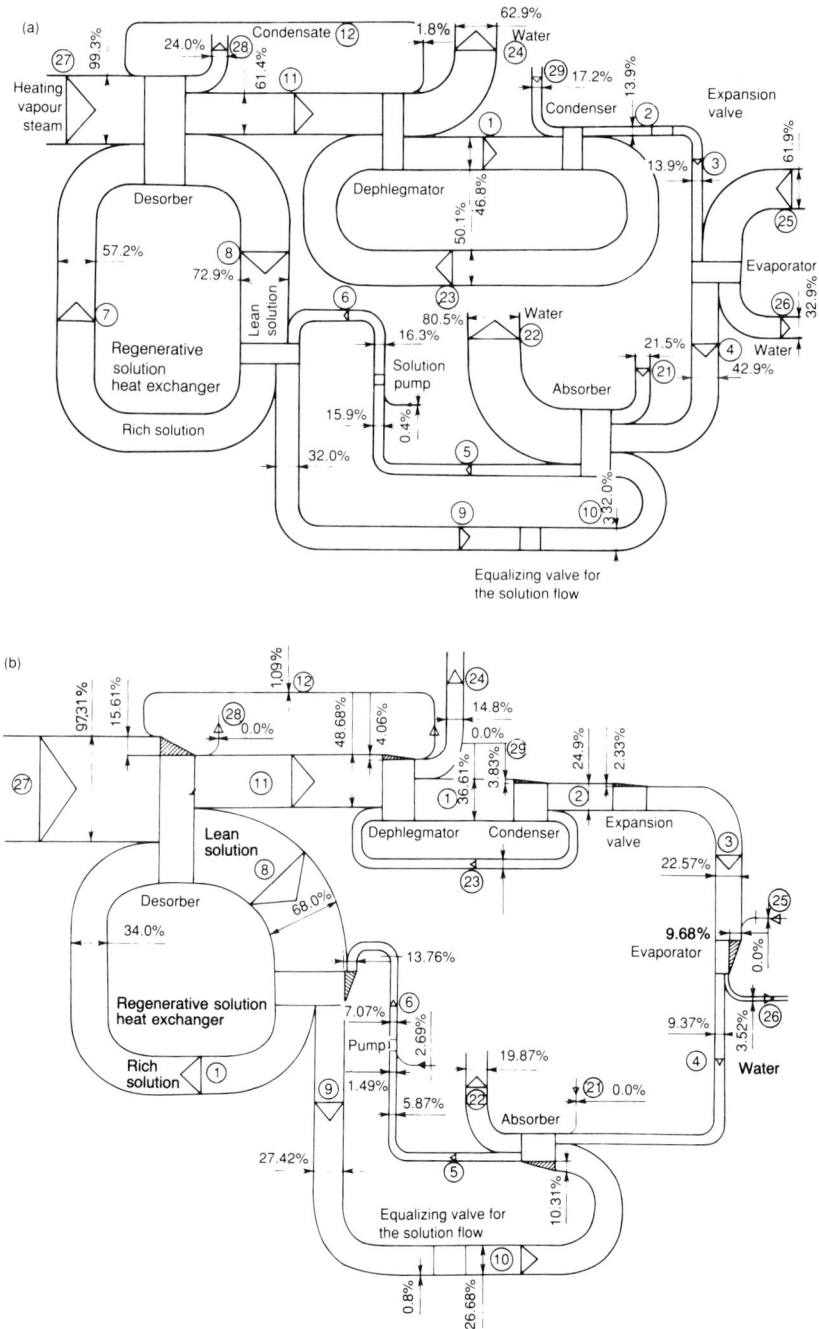

Figure 5.14   *Flow diagrams: (a) energy fluxes; (b) exergy fluxes*

Table 5.4 *Parameters characterizing the absorption cycle heat pump*

(a) Working fluid

No.	Fluid	T (°C)	P (MPa)	z (kg/kg)	i (kJ/kg)	s (kJ/(kg K))	b (kJ/kg)	m^* (kg/kg NH_3)	I/E_D (%)	B/B_D (%)
1	Vapour	82.69	2.020	0.9900	1753.60	4.674	338.47	1.000	46.8	36.61
2	Condensate	38.75	2.000	0.9900	521.90	0.647	230.18	1.000	13.9	24.90
3	Liquid–vapour $x = 16.7\%$	−6.58	0.329	0.9900	521.30	0.715	208.63	1.000	13.9	22.57
4	Wet vapour $x = 97.5\%$	11.08	0.329	0.9900	1608.60	4.839	86.60	1.000	42.9	9.37
5	Rich solution	46.86	0.326	0.3585	107.70	0.530	9.79	5.538	15.9	5.87
6	Rich solution	47.16	2.040	0.3585	110.39	0.533	11.81	5.538	16.3	7.07
7	Rich solution	107.34	2.040	0.3585	387.40	1.324	56.75	5.538	57.2	34.00
8	Lean solution	153.00	2.040	0.2194	602.60	1.900	138.73	4.538	72.9	68.11
9	Lean solution	78.20	2.040	0.2194	264.50	1.029	55.85	4.538	32.0	27.42
10	Lean solution	79.06	0.326	0.2194	264.50	1.035	54.23	4.538	32.0	26.68
11	Vapour	123.68	2.020	0.9186	1917.50	4.972	374.62	1.201	61.4	48.68
12	Reflux	82.69	2.020	0.5640	340.60	1.008	50.26	0.201	1.8	1.09

(b) Substances of the source

No.	Fluid	T (°C)	P (MPa)	i (kJ/kg)	s (kJ/(kg K))	b (kJ/kg)	m^* (kg/kg NH_3)	I/E_D (%)	B/B_D (%)
21	Water	20.2	0.400	83.8	0.299	0.00	9.646	21.5	0.00
22	Water	74.6	0.400	312.6	1.015	18.98	9.646	80.5	19.87
29	Water	20.0	0.400	83.8	0.299	0.00	7.770	17.2	0.00
23	Water	57.7	0.400	241.8	0.807	9.38	7.770	50.1	7.88
24	Water	72.5	0.400	303.8	0.990	17.61	7.770	62.9	14.80
25	Water	15.0	0.400	62.9	0.227	0.00	36.892	61.9	0.00
26	Water	8.0	0.400	33.5	0.123	0.88	36.892	32.9	3.52
27	Heating steam	158.0	0.588	2757.0	6.786	665.70	1.351	99.3	97.31
28	Condensate	158.0	0.588	667.0	1.928	0.00	1.351	24.0	0.00

In Tables 5.4 and 5.5 all quantities are referred to 1 kg of the working fluid vapour.
For the simplicity of calculations the exergy at points: 21, 25, 28 and 29 was assumed to be zero.
*Driving energy $E_D = m_{27}^* i_{27} + m_5^* (i_6 - i_5)/(\eta_{m,sr}\eta_{el})$
*Driving exergy $B_D = m_{27}^* b_{27} + m_5^* (b_6 - b_5)/(\eta_{m,sr}\eta_{el})$

Table 5.5 *Characteristics of plant. Exergy losses are referred to the driving exergy*

Apparatus	AU (kJ/kg K))	η	ΔT_{in} (°C)	q (kJ/kg)	Exergy loss (%)
Condenser	56.76	–	21.70	1231.9	3.83
Evaporator	133.78	–	8.12	1086.7	9.68
Absorber	177.57	0.9	12.46	2212.7	10.32
Regenerative solution heat exchanger	40.14	–	38.22	1534.1	13.76
Desorber	143.24	0.9	19.72	2824.1	15.62
Dephlegmator	29.19	–	16.49	481.4	4.06
Expansion valve	–	–	–	–	2.33
Throttling valve for the solution	–	–	–	–	0.80
Solution pump	–	0.6	–	–	1.49

Table 5.6 *Values of parameters of the subset $\mathscr{A}_{a3}$ for the heat transformer*

Cause of origin	Parameter
Heat transfer	Temperature drops of the source substances $\delta T_{\text{de}} = 10\ \text{K}, \delta T_{\text{ev}} = 5\ \text{K},$ $\delta T_{\text{co}} = 3\ \text{K}$
	Minimum temperature differences at the exchanger inlets or outlets $\Delta T_{\text{ab}} = 4\ \text{K}, \Delta T_{\text{de}} = 4\ \text{K},$ $\Delta T_{\text{ev}} = 3\ \text{K}, \Delta T_{\text{co}} = 2\ \text{K},$ $\Delta T_{\text{se}} = 5\ \text{K}$
Mass transfer	$\eta_{\text{ab}} = 0.9,$ $\eta_{\text{de}} = q_{\text{dp}}/q_{\text{dp, min}} = 1.2,$ $q_{\text{dp, min}} = $ for an infinite number of trays
Resistance to flow	$\delta P_{\text{ls}} = \dfrac{P_{\text{de}} - P_{\text{co}}}{P_{\text{co}}} = 0.025,$ $\delta P_{\text{hs}} = \dfrac{P_{\text{ev}} - P_{\text{ab}}}{P_{\text{ev}}} = 0.025,$ $\eta_{\text{m,sl}} = 0.6; \eta_{\text{m,sr}} = 0.6$

for energy and exergy, and Table 5.5 gives the values of characteristic parameters in the nodes of the installation and the exergy losses.

The calculations included an analysis of the effect of parameters characterizing the heat and mass transfer kinetics and exergy losses on the COP and exergetic efficiency. Detailed results are shown in Figures 5.15, 5.16, 5.17 and 5.18. It is

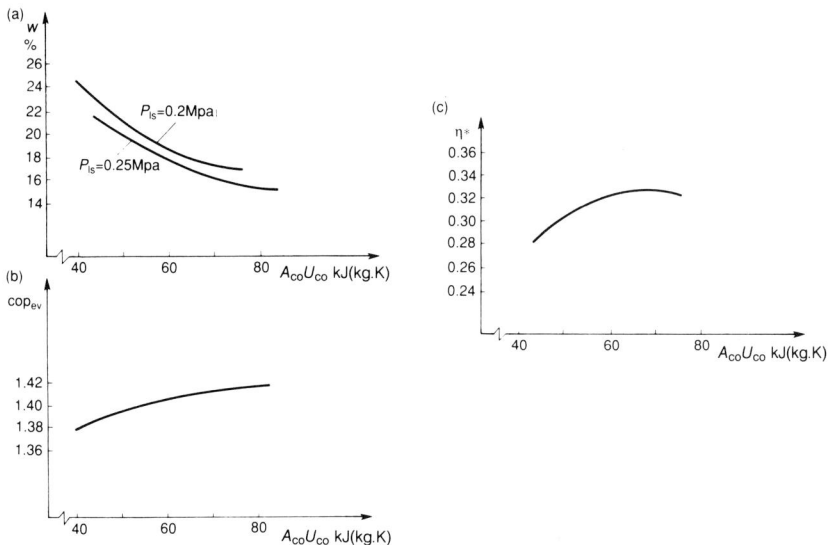

Figure 5.15   *Effect of heat transfer conditions in the condenser on:* (a) *dryness fraction of the two-phase mixture;* (b) *coefficient of performance;* (c) *exergetic efficiency*

worth mentioning that a reduction of exergy losses in the condenser and expansion valve causes an increase of these losses in the evaporator and thus the relationship $\eta^*_{aa} = d(AU)_{co}$ has an optimum value.

Humid air ($w = 1$) at temperature T_z provides the source of waste heat for the heat transformer. This heat is transferred in the desorber and evaporator, mainly as a result of the condensation of water vapour (contained in air). A part of this heat is used to generate medium-pressure steam, and the rest is transferred in the condenser and dephlegmator of the rectifier column. Table 5.6 shows the values of parameters (subsets $\mathscr{A}_{a2}$ and $\mathscr{A}_{a3}$) characterizing the exergy losses. The effect of external parameters (temperatures) on the COP and exergetic efficiency is shown in Figure 5.19. Two regimes of variations of these coefficients can be distinguished. In the first regime the COP decreases very slightly with an increase of temperature in the absorber, and the exergetic efficiency increases. These variations can be determined on the basis of the theoretical cycle discussed earlier. In the second regime both coefficients decrease rapidly with the increase of the absorber temperature, which is evidence of the importance of exergy losses in the approximate cycle. Temperature in the absorber increases in parallel with the decrease of temperature in the condenser, which also means the decrease of condenser pressure. On one hand this causes an increase of the enthalpy of condensation, and thus the quantity of heat release to the environment, and on the other hand, a decrease of the fraction of working fluid in the lean solution, which leads to an increase in the thermal load in the absorber. A flow diagram of exergy losses for the assumed temperatures of the sources is shown in Figure 5.20. Because

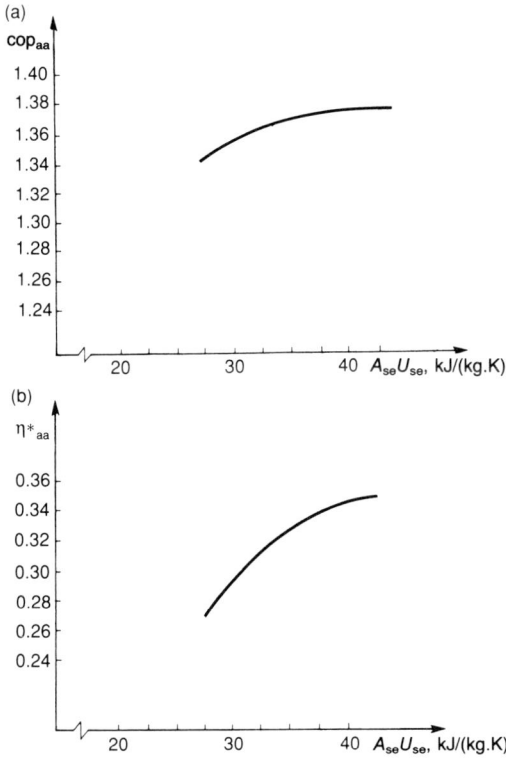

Figure 5.16   *Effect of heat transfer conditions in the regenerative solution heat exchanger on:* (a) *coefficient of performance;* (b) *exergetic efficiency*

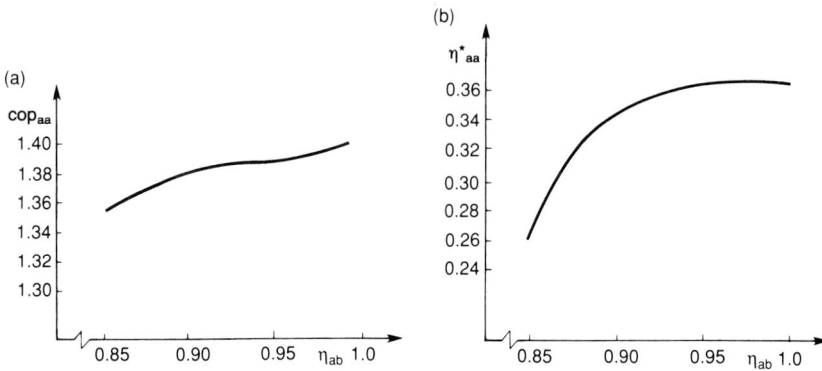

Figure 5.17   *Effect of mass transfer conditions in the absorber on:* (a) *coefficient of performance;* (b) *exergetic efficiency*

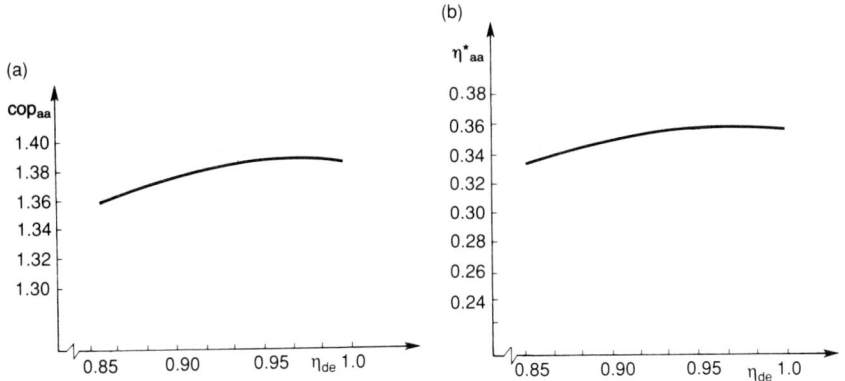

Figure 5.18   *Effect of mass transfer conditions in the desorber node on:* (a) *coefficient of performance;* (b) *exergetic efficiency*

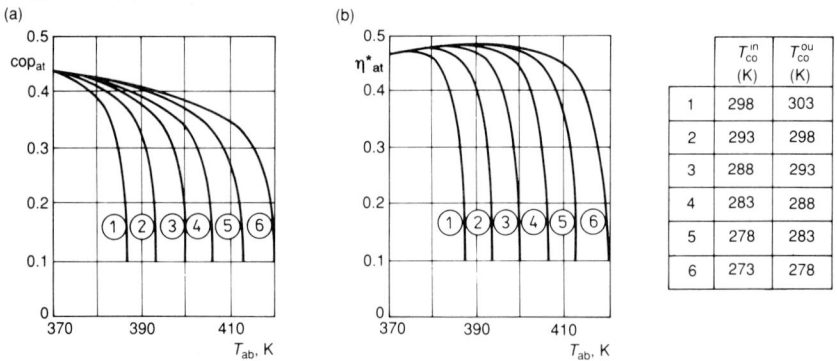

	T_{co}^{in} (K)	T_{co}^{ou} (K)
1	298	303
2	293	298
3	288	293
4	283	288
5	278	283
6	273	278

Figure 5.19   *Coefficients:* (a) *coefficient of performance;* (b) *exergetic efficiency of the approximate cycle of a heat transformer*

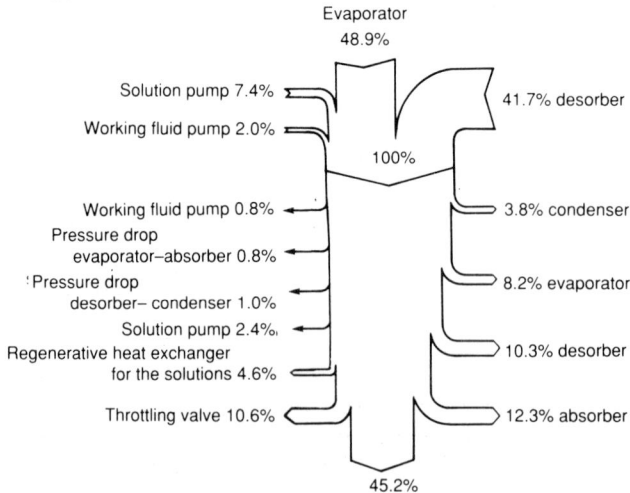

Figure 5.20   *Flow diagram of the exergy losses in a heat transformer*

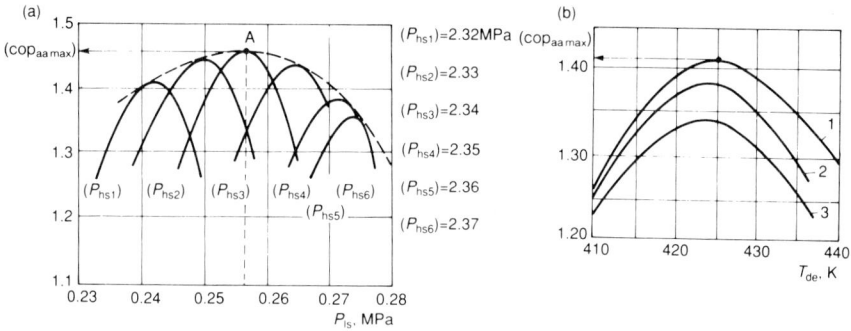

Figure 5.21 *Results of optimization calculations for an absorption cycle heat pump using a pair R717–water: (a) effect of pressure on the value of coefficient of performance; (b) effect of the efficiency of mass transfer in the absorber and desorber on maximum values of the coefficients of performance. (1)* $\eta_{ab} = 1.0$, $\eta_{de} = 1.0$; *(2)* $\eta_{ab} = 1.0$, $\eta_{de} = 0.7$; *(3)* $\eta_{ab} = 0.7$, $\eta_{de} = 0.7$

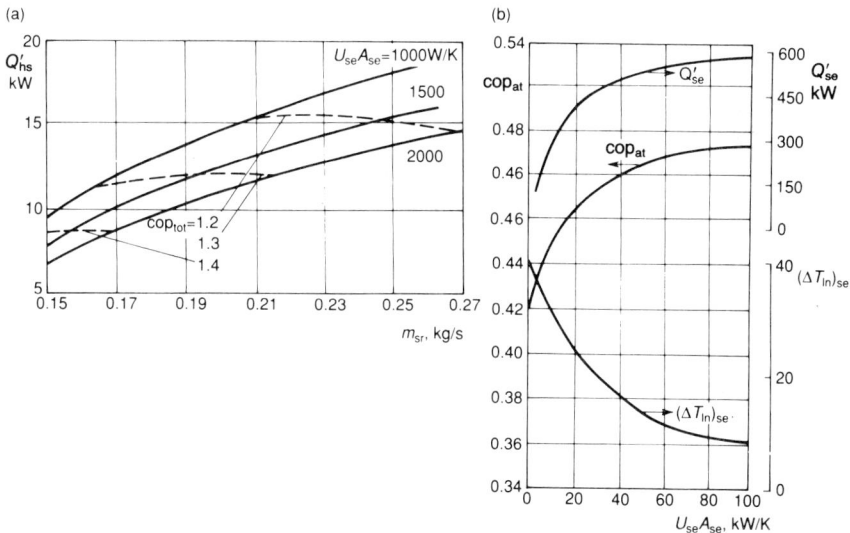

Figure 5.22 *Effect of the conditions of heat transfer in the regenerative heat exchanger for solutions on the heat flux* Q'_{hs}: *(a) for an absorption cycle heat pump using the R22–E181 pair [43]; (b) for a heat transformer using the R718–lithium bromide pair [69]*

the mass flow rate of air directed to the evaporator is larger than that directed to the desorber, the exergy flux delivered to the evaporator is larger than that delivered to the desorber. The exergy losses, however, are larger in the desorber than in the evaporator, which is the result of the rectification of vapours in the column. Large losses in the throttling valve result from the pressure difference ($\Delta P = 3.1$ MPa) between the absorber and desorber.

The aim of introducing this method of optimization was to show its general advantages in analysing exergy losses of heat pumps. Figures 5.21 and 5.22 show the results of analyses by Baehr [59], Sano [69] and Trommelmans [44]. With calculations of this type there is no more need to use the comparative cycles. The calculations, however, require the use of quite complex computing techniques, and thus comparative cycles are useful for a preliminary assessment.

Chapter 6

Examples

The examples shown in this chapter illustrate the methods of calculation of theoretical cycles and exchanger areas of heat pump installations. The selection of examples was motivated by aiming at a discussion of practically implemented heat pumps and indicating their potential applications (Table 6.1). For this reason each example is different in its formulation, range of calculations and the degree of detail in the presented information. Examples of applications of heat pumps in heating are given in, among others: [70], [32], [71], [72], [73], [74], [75], [76] and [77], and their applications in industry in: [78], [79], [80], [81], [82], [83], [84], [85], [86], [87], [88], [49], [89], [90], [91], [92], [92], [94], [95], [96], [97], [69], [98], [99], [100], [101] and [102]. The possibilities of manufacturing heat pumps in Poland and their potential applications are shown in: [103], [104], [105], [106], [107], [108], [109], [110], [111], [112], [113] and [114].

In the book we have not included, because of climatic conditions prevailing in Poland, any heat pumps utilizing solar energy in their low temperature source. Examples of such applications are given among others in: [115], [116], [117] and [118]. Table 6.1 lists the examples discussed in this chapter and gives their brief characteristics.

Calculations of the cycles were based on the methods discussed in Chapters 4 and 5, and the surface areas of exchangers were determined in a simplified manner; data necessary for their design can be found in the literature on heat and mass exchanger calculations. As the book centres on issues of thermodynamics, design calculations are omitted in the examples. These calculations, particularly for the compressors, are very specific and for this reason they are limited here to the determination of power necessary to drive the compressor at an assumed efficiency. Design calculations of exchangers, because of the range of pressures and temperatures encountered here, are of quite standard character and do not cause additional problems.

Example 6.1

This example illustrates the application of heat pumps utilizing the energy of atmospheric air in the heating of individual houses, for various levels of power, various types of heat pump, working fluids employed, range of calculations etc. Example 6.1.1 shows a method of selecting a mechanical vapour compression heat pump driven by an electric motor. In example 6.1.2 we discuss the reduction of exergy losses by using an azeotropic mixture as the working fluid, and Example 6.1.3 illustrates the use of solar energy collectors and heat accumulators as

Table 6.1 Characteristics of the examples

Example number	Type of heat pump	Main task	Low-temperature heat source	High-temperature heat source (kW)	Working fluid or pair	Example illustrates
6.1		Individual house heating	Atmospheric air			
6.1.1	Mechanical vapour compression			3.3	R22	Selection of heat pump from a catalogue
6.1.2	Mechanical vapour compression				Zeotropic mixture	Reduction of exergy losses
6.1.3	Mechanical vapour compression				R22	Utilization of solar energy
6.1.4	Absorption			20	R717–water	Design calculations of the cycle and exchangers
6.1.5	Absorption				R22–E181	
6.2		Heating a greenhouse of arable area approx. 700 m²	Ground water			
6.2.1	Variant I mechanical compression			670	R12	Comparison of mechanical vapour compression and resorption heat pumps
6.2.2	Variant II resorption				R717–water	
6.3	Ejector	Heating of technological water	Water from a lake	83 600	R11	Calculation of a theoretical cycle taking into account losses in the ejector and valve; analysis of exergy losses in the cycle
6.4		Reversing the heat flows in a rectification column	Heat of condensation (from the dephlegmator of a rectification column)			Implemented and tested possibility of application of a screw compressor
6.4.1	Absorption			580	R718–LiBr	
6.4.2	Heat transformer			2350	R718–LiBr	
6.4.3	Mechanical compression			1965	R718	

Table 6.1 continued

Example number	Type of heat pump	Main task	Low-temperature heat source	High-temperature heat source (kW)	Working fluid or pair	Example illustrates
6.5	Mechanical compression	Reversing the heat flows in a sublimation dryer (freeze-dryer)	Water from a condenser	130 kJ/kg	R12	Possibility of using a heat pump co-operating with a refrigeration plant
6.6 6.6.1 6.6.2	Mechanical vapour compression	Compression and re-utilization of vapours: – from a brewing vat in a brewery – from an evaporator battery in sugar refinery	Heat of condensation of vapours generated in technological process, employed also as the working fluid	6470 6470	R718	Implemented and tested application of thermo-compression in industrial plant
6.7	Mechanical vapour compression	Heating of water in a bottle-washing plant	Waste water from the washing plant	240	Blend of R12 and R22	Implemented and tested recovery of waste heat by returning it to the installation
6.8 6.8.1	Mechanical vapour compression	Central heating of a large housing estate	Technological steam from sugar refinery purified waste water	18 900		Implemented and tested cooperation of a heat pump with large heating systems
6.8.2	Mechanical vapour compression		Technological water from cooling flue gases generated in a waste incinerator	3300	R12	
6.8.3	Absorption			55 000	R12 R718–lithium bromide	
6.9	Mechanical vapour compression	Energy economy in an industrial installation	Industrial medium to be cooled	150	R123	Use of 'pinch point' technology to determine the best location of heat pump in the plant diagram

additional low temperature sources Examples 6.1.4 and 6.1.5 illustrate the design calculations of an absorption cycle heat pump having power relatively high for the demand of a single family home (20 kW), supplying the domestic utility hot water and heating the space via a hot water central heating installation. These absorption cycle heat pumps differ in the pairs of the working fluid–sorbent employed (R717–water, R22–E181). Calculations were performed for: $T_{hs} = 333 \, K$, $T_{ls} = 285 \, K$, $T_{water, inlet} = 313 \, K$ and temperature differences at the exchanger outlets not smaller than 5 K.

Example 6.1.1

Mitsubishi Heavy Industries Ltd and Klima Ltd offer, in the Polish market a range of the 'Miniman' mechanical vapour compression electrically driven heat pumps, for use in heating of individual houses. Heat is absorbed from the external air, and air heated in the condenser, circulating inside the building, is distributed through ventilating ducts. Commercial literature distributed by the company includes data in numerical and graphical form. The numerical data, parts of which are quoted here (external dimensions and weights of individual plant are omitted), refer to an external temperature of 279 K and internal temperature 293 K. Figure 6.1.1 shows the data on thermal power, drive power and COP_{rc} as a function of the external temperature, for varying degrees of icing on the evaporator (icing appears already at the external temperature of 278 K).

Selection of the heat pump and analysis of the build series was done on the assumption that it would be used to heat a bungalow in the shape of a cuboid with dimensions $10 \times 15 \times 3 \, m$ and characterized by the overall heat transfer coefficient of $0.3 \, W/(m^2 \, K)$ at nominal conditions. The analysis of heat pump operation was also performed for other conditions. Due to a lack of detailed data in the catalogue, many parameters are calculated. The results shown in Table 6.1.1 are compiled for the following variants: variant A: external temperature 279 K, internal temperature 293 K, due to the power surplus this heat pump operates only 17.5 hours in every 24 hours ($2.4 \, kW/3.3 \, kW = 0.73$, $0.73 \times 24 \, h = 17.5 \, h$); variant B–external temperature 268 K, internal temperature 293 K, additional electric heater; variant C–external temperature 268 K, internal temperature is set at the level of 283 K.

Table 6.1.1   *Characteristics of the Miniman range of heat pumps*

Miniman type	Thermal power in the high-temperature heat source (kW)	Maximum flow of heated air (m³/h)	Noise emission (dB)
I	2.4	480[a]	44–48
II	3.3	480	44–49
III	4.2	540	50–53

[a]Step variations of 10% and 20%.

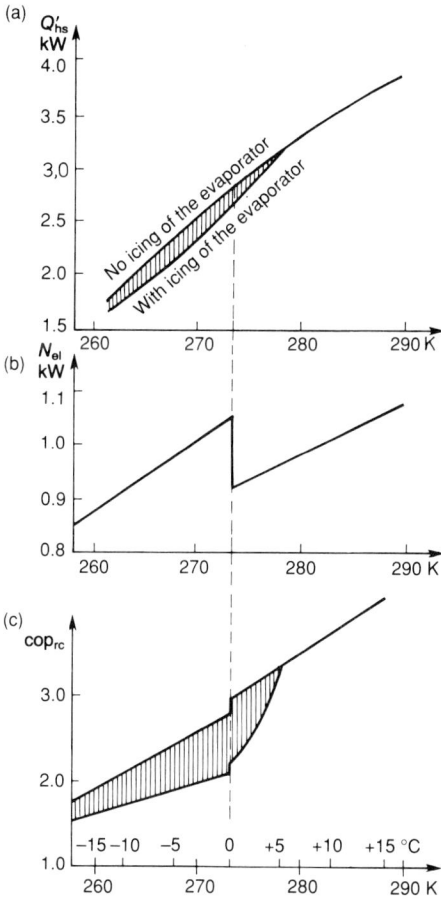

Figure 6.1.1 *Characteristics of a Miniman II heat pump as a function of external temperature, for temperature of the interior 293 K:* (a) *thermal power in the high-temperature heat source;* (b) *drive power (the jump is caused by switching on of the evaporator defrosting system);* (c) *COP$_{rc}$ of a real heat pump*

The power demand at the high-temperature source level is determined by the magnitude of heat flux exchanged between the building and the environment, and by the energy needed to heat the flow of the exchanged air, estimated at 20% of the circulating air. The exchanged heat flux is calculated for the surface area of the building: walls, $2[(10 \times 3) + 15 \times 3)] = 150\,\text{m}^2$; floor and ceiling $2(10 \times 15) = 300\,\text{m}^2$; total $450\,\text{m}^2$. The exchanged heat flux for the variant A is:

$$50 \times 0.3 \times (20 - 6) = 1.89\,\text{kW}$$

The energy needed to heat the air (taken in from the outside at the rate equal to 20% of the circulating air) is:

$(480/3600) \times 0.2 \times 1.005 \times (20 - 6) = 0.375\,\text{kW}$

The total power demand at the high temperature source level is:

$Q'_{hs} = 1.89 + 0.375 = 2.265\,\text{kW}$

It was decided to use the Miniman II heat pump, despite the fact that it will not satisfy the demand at the lower external temperatures. At these conditions either the internal temperature will settle at a level much lower than required, or additional heaters need to be switched on, or alternatively heating must be limited to part of the building only.

The results obtained for all three variants are listed in Table 6.1.2.

Table 6.1.2   *Parameters of a Miniman heat pump*

Characteristic parameters	Variants		
	A	B	C
Power demand for heating the building (kW)	2.30	4.20	2.40
Thermal power of the Miniman II heat pump in the high-temperature heat source (kW)	3.30	2.40	2.40
Electrical drive power (kW)	0.98[a]	0.99	0.86
Auxiliary heater power (kW)	–	1.80	–
External temperature (K)	279.00	268.00	268.00
Internal temperature (K)	293.00	293.00	283.00
Condensation temperature (K)	321.00	321.00	313.00
Evaporation temperature (K)	268.00	257.00	257.00
Temperature of air leaving the condenser, heating the interior (K)	317.00	317.00	306.00

[a]Compressor drive power; omitted is the power to drive the fans located in the condenser and evaporator.

It is appropriate to mention the simplicity of the plant and its installation, e.g. in comparison with a heat pump heating the house using hot water. Hence its low cost, at the level of 1200 pounds sterling. The low price is an advantage of this solution, but its parameters are limited and it does not provide the possibility of supplying the utility hot water.

Example 6.1.2

In a vapour compression cycle Miniman III heat pump the standard R22 working fluid is replaced by a binary zeotropic mixture. This mixture condenses in non-isothermal conditions, which makes it possible to reduce exergy losses in the condenser. For this heat pump and for the following conditions, good properties are shown by a mixture of 0.35 R134 and 0.65 R32: inlet air temperature to condenser 21°C, evaporator 7°C, subcooling of condensate 5°C, non-superheating of steam in the evaporator. The non-isothermal nature of the process of condensation of the zeotropic mixture is characterized by the change in temperature along the length of the condenser, as shown in Figure 6.1.2. The approximate cycle realized alternatively by the R22 and the zeotropic mixture is shown in Figure 6.1.3, and the corresponding intensive parameters are listed in Tables 6.1.3 and 6.1.4.

The exergy losses are shown as a flow diagram in Figure 6.1.4, and their numerical values are listed in Table 6.1.5.

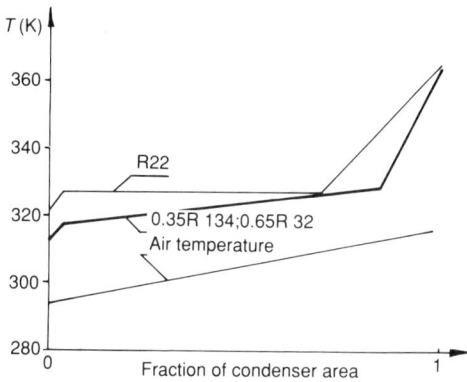

Figure 6.1.2 *Temperature changes of the zeotropic mixture along the condenser length; for comparison, the changes of R22 working fluid are also shown*

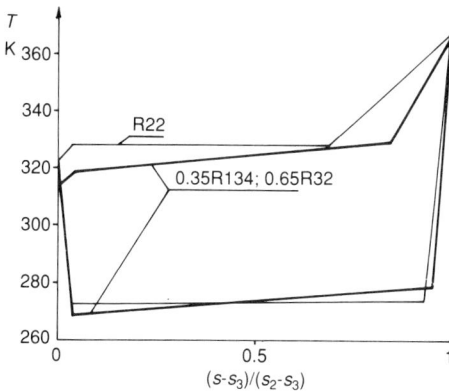

Figure 6.1.3 *Approximate cycle realized by the zeotropic mixture. The dimensionless entropy is introduced to enable comparison of this cycle with the alternative cycle realized using R22*

Table 6.1.3 *Intensive parameters of the approximate cycle realized with the R22 working fluid*

	T (K)	P (MPa)	i (kJ/kg)	s (kJ/kg K)
1	272.4	0.479	401.84	1.748
2	366.4	2.167	451.7	1.788
2'	327.2	2.167	416.2	1.600
3	327.2	2.167	262.3	1.209
3'	322.2	2.167	254.6	1.187
4	272.4	0.479	254.6	1.209

Table 6.1.4 *Intensive parameters of the approximate cycle realized with the zeotropic mixture (0.35 R134 + 0.65 R32)*

	T (K)	P (MPa)	i (kJ/kg)	s (kJ/kg K)
1	277.2	0.492	478.5	2.132
2	365.2	2.073	542.9	2.171
2'	328.6	2.073	498.5	2.043
3	318.0	2.073	282.6	1.372
3'	313.0	2.073	271.8	1.338
4	268.2	0.492	271.8	1.367

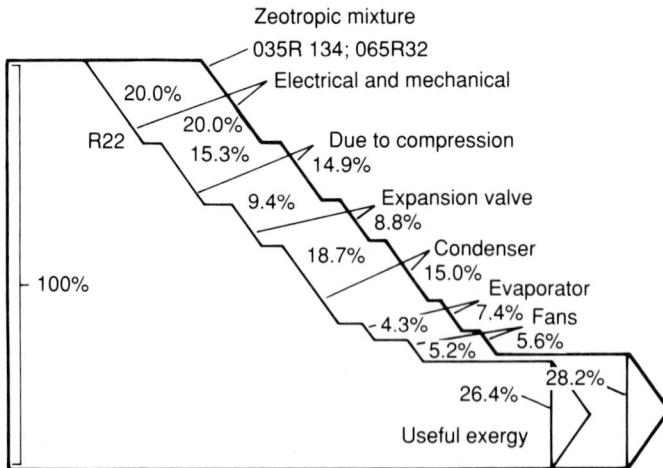

Figure 6.1.4   *Exergy flow diagram illustrating the percentage losses of exergy in both cycles*

Table 6.1.5 *Exergy losses in approximate cycles of Figure 6.1.4 (including the conversion of electrical to mechanical energy)*

Exergy losses	R22	Mixture 0.35 R134, 0.65 R32	Gain due to mixture
Electrical and mechanical	280	256	24
Due to compression	223	202	21
Expansion valve	131	115	16
Condenser	262	197	65
Evaporator	60	97	−37
Fans	74	76	−2
			87

Example 6.1.3

A vapour compression cycle heat pump Miniman I is used to heat a house. Its lower temperature source utilizes solar energy, as shown in Figure 6.1.5. The concept is realized by using a flat solar collector with air flowing through, and an energy collector comprising a bundle of tubes filled with paraffin wax and sealed at both ends [119]. These tubes are arranged transversely to the air flow.

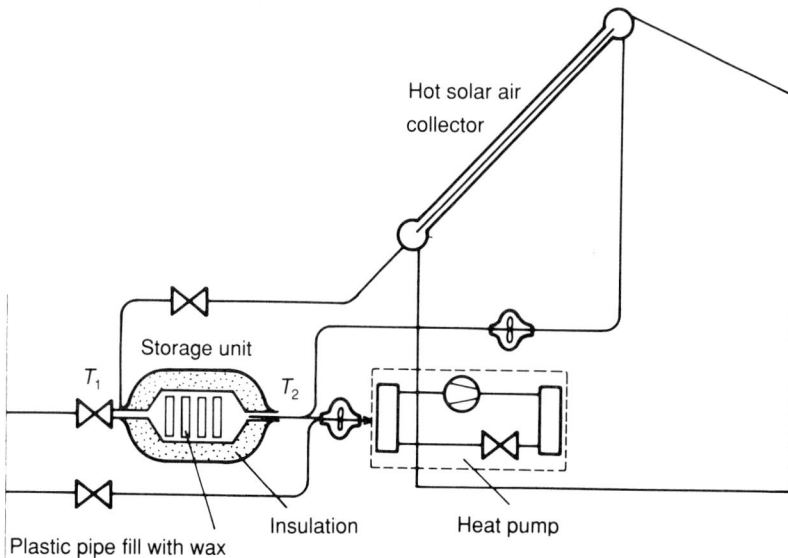

Figure 6.1.5 *Diagram of an installation comprising a solar collector and an energy storage accumulator boosting the low-temperature heat source of a vapour compression cycle heat pump; energy extracted from the solar collector is temporarily stored in the accumulator*

During the period of maximum insolation, 5–6 h per day, energy is extracted from the collector by a separate air circuit and stored in the accumulator. During this period the evaporator of the heat pump is heated by an air stream taken directly from the atmosphere ($480 \, m^3 h^{-1}$). For the remainder of the day the evaporator air stream is first passed through the energy storage accumulator, where it is pre-heated.

Simulation calculations led to a proposed collector consisting of 10 elements having total area of $55 \, m^2$ (each element has identical external dimensions of $5.5 \times 1 \, m$ and a gap of 1 cm). Calculations were performed for a collector covered with a single plate of glass, and the absorptivity of its bottom was assumed to be 0.9. The energy storage accumulator was proposed as an array of tubes of 50 mm diameter. Twenty-five of these tubes were placed transversely to flow, and 40 along the length (a total of 1000 tubes). The total mass of wax filling the tubes is 1600 kg, and the external dimensions of the proposed energy storage accumulator are $0.8 \times 1.75 \times 2.4 \, m$.

The calculated increases in air temperature resulting from its pre-heating in the energy storage accumulator are shown in Figure 6.1.6. The calculations are based on monthly average temperatures and an average value of insolation energy. The transients characterizing the daily changes in heat transfer conditions were neglected. This simplification can be treated in two ways: either the calculations are performed for daily average conditions or the air flow rate is controlled so that its temperature increase remains constant.

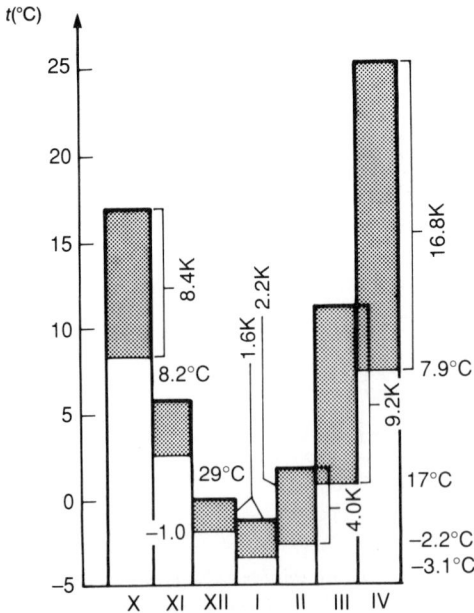

Figure 6.1.6 *Increases in air temperature resulting from its pre-heating in the energy storage accumulator*

Temperature increases below 3°C cannot be utilized in practice, so that during the winter (from November to January) solar energy cannot be used to supplement the low-temperature heat source. During the remaining months pre-heating of air results in higher values of the COP_{rc}, which are shown in Figure 6.1.7 for the Miniman I pump (calculations take into account the effect of icing occurring below 5°C).

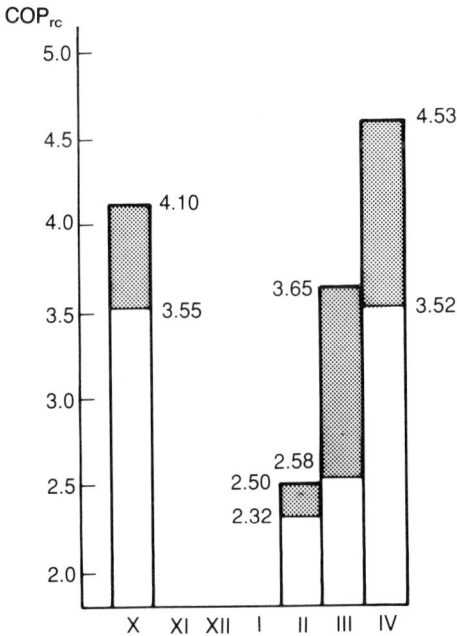

Figure 6.1.7 *Values of the coefficient of performance of the vapour compression cycle heat pump boosted by the solar collector; nominal values of COP are also given*

Example 6.1.4

The diagram of installation is shown in Figure 6.1.8, and its cycle in Figure 6.1.9.

Calculations of the approximate cycle

Calculations were performed following the algorithm introduced in Chapter 5, making it possible to determine the optimum parameters. The calculated extensive and intensive parameters are shown in Table 6.1.6, and the values of exchanged heat fluxes in Table 6.1.6. The calculated power necessary to drive the solution pump is $N_p = 0.06\,kW$ and the numbers of (mass) transfer units in the parts of the

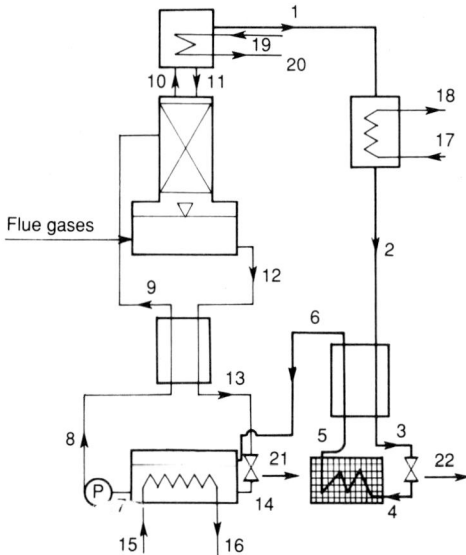

Figure 6.1.8   *Diagram of an absorption cycle heat pump employing the R717–water pair of working fluids*

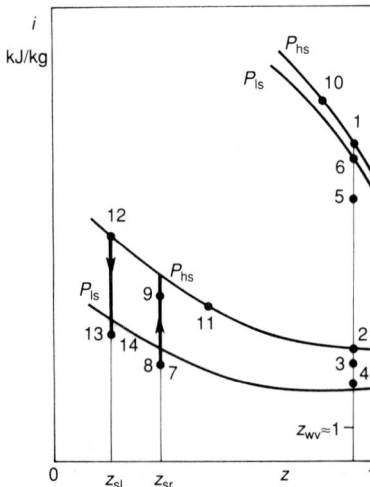

Figure 6.1.9   *Thermodynamic cycle of heat pump from Figure 6.1.8, in the i–z coordinates*

column above and below the feed point are correspondingly equal: $NTU_1 = 1.4$ and $NTU_2 = 0.6$. The COP_{aa} is equal to:

$$COP_{aa} = (Q'_{ab} + Q'_{dp} + Q'_{co})/(Q'_{de} + N_p) = (9.72 + 4.38 + 5.67)/(14.24 + 0.06) = 1.37$$

Table 6.1.6 *Listing of extensive and intensive parameters calculated in Example 6.1.4 (point numbers correspond to points in Figure 6.1.8)*

Point number	Temperature T	Pressure (MPa)	Mass fraction	Mass flow (kg/s)	Enthalpy (kJ/kg)
1	353	2.340	0.991	0.005	1392.30
2	328				259.10
3	291				24.30
4	270	0.253			as above
5	263				1154.20
6	323				1389.00
7	320		0.342	0.022	−21.30
8	320	2.340			−18.30
9	403				389.50
10	401		0.885	0.007	1650.30
11	353		0.626	0.002	150.40
12	448		0.155	0.017	665.20
13	331				139.80
14	333	0.253			as above
15	313	0.100	Water	0.181	167.52
16	326				221.22
17	313			0.039	167.52
18	338			As above	312.65
19	313			0.019	167.52
20	368				398.05
21	285		Air	0.710	10.22
22	275		As above		2.10

Table 6.1.7 *Calculated heat fluxes*

Part of installation	Heat flux (kW)	Comments
Absorber	9.72	High-temperature heat source
Dephlegmator	4.38	High-temperature heat source
Condenser	5.67	High-temperature heat source
Evaporator	5.77	Low-temperature heat source
Desorber	14.24	External heat source (heat from burning oil)
Regenerative vapour heat exchanger	1.17	–
Regenerative solution heat exchanger	8.97	–

Exchanger calculations

Absorber. It was assumed that the lean solution flows down as a liquid film (a few millimetres thick) over the external surface of the coil, with water flowing inside. This coil is located in a tank filled with ammonia vapour. The surface area of the coil was calculated for the following assumptions: mass transfer coefficient $\beta = 0.10 \, \text{kg/(m}^2\text{s})$ and the overall heat transfer coefficient $U_{ab} = 800 \, \text{W/(m}^2\text{K})$. The driving force: for mass transfer is $\Delta z_{ln} = 0.112$ and for the heat transfer is $\Delta T_{ln} = 5.95 \, \text{K}$. Hence the necessary transfer area: for mass transfer is $A_{ab} = m_{wv}/(\beta \Delta z_{ln}) = 0.398 \, \text{m}^2$ and for the heat transfer is $A_{ab} = Q'_{ab}/(U_{ab} \, \Delta T_{ln}) = 2.15 \, \text{m}^2$. Thus for further calculations the surface area was assumed as $A_{ab} = 2.15 \, \text{m}^2$.

Dephlegmator. A shell and tube exchanger design with horizontal tubes was assumed. Vapours flowing in the shell area partly condense, releasing the heat, which is transferred to water flowing in the tubes. The overall coefficient of heat transfer is assumed to be $U_{dp} = 1000 \, \text{W/(m}^2 \, \text{K})$, and the value of the driving force as $\Delta T_{ln} = 36.3 \, \text{K}$ (according to data in Table 6.1.1). The area of the exchange surface of the dephlegmator $A_{dp} = Q'_{ab}/(U_{ab} \, \Delta T_{ln}) = 0.12 \, \text{m}^2$.

Condenser. The adopted design is the same as that of the dephlegmator. The magnitude of the driving force (based on the data in Table 6.1.1) $\Delta T_{ln} = 6.45 \, \text{K}$, and assuming $U_{co} = 1000 \, \text{W/(m}^2 \, \text{K})$, the surface area of the condenser $A_{co} = Q'_{co}/(U_{co} \, \Delta T_{ln}) = 0.84 \, \text{m}^2$.

Evaporator. A lamelle heat exchanger design was adopted. For such design $U_{ev} = 30 \, \text{W/(m}^2 \, \text{K})$ (referring to $1 \, \text{m}^2$ of the lamelle surface area). The driving force $\Delta T_{ln} = 8.93 \, \text{K}$, and thus the area of evaporator $A_{ev} = Q'_{ev}/(U_{ev} \, \Delta T_{ln}) = 23.88 \, \text{m}^2$.

Desorber. This consists of a boiler and rectification column. The boiler is a cylindrical tank heated with an oil burner. It is assumed that inside the boiler the liquid boils in a large volume, and is subject to gravity forces. The heat transfer coefficient is $\alpha = 2.4 \times 104 \, \text{W/(m}^2 \, \text{K})$; it was further assumed that the temperature of the inside wall of the tank is $5 \, \text{K}$ higher than the temperature of the boiling lean solution, that is $T_{w,int} = 453 \, \text{K}$. Thus the surface area of the desorber $A_{de} = Q'_{de}/(\alpha \, \Delta T) = 0.121 \, \text{m}^2$. For this surface area, the surface density of heat flux is $q_{de} = Q'_{de}/A_{de} = 1.17 \times 10^5 \, \text{W/m}^2$ and is smaller than the critical value $q'_{cr} = 3.4 \, 10^6 \, \text{W/m}^2$. Assuming the thickness of the boiler wall $\delta w = 5 \times 10^{-3} \, \text{m}$ and its material as St35X with thermal conductivity $k = 52.33 \, \text{W/(mK)}$, the temperature of the external wall $T_{w,ex} = T_{w,int} + Q_{de} \, \delta w/(A_{de}k) = 464.17 \, \text{K}$.

Rectification column. This was calculated assuming that its diameter is $D = 0.2 \, \text{m}$ and that it is filled with Raschig rings of diameter $d = 0.0123 \, \text{m}$. The value of the mass transfer coefficient for such rings is $\beta = 0.6 \, \text{kg/(m}^2\text{s})$. The total height of the column was calculated using the equation: $H = \text{HTU} \times \text{NTU}$, where $\text{HTU} = m_{sr}/(\beta \pi D^2/4)$. The value of mass flux, averaged over the length of column, is $m_{sr} = 0.01 \, \text{kg/s}$ and thus $H = 0.95 \, \text{m}$. The height at which the rich solution is supplied, as measured from the bottom of the column, is $H_1 = \text{HTU} \times \text{NTU}_1 = 0.282 \, \text{m}$.

Regenerative heat exchanger for the vapours. A concentric 'tube in tube' design was adopted, with the outer surface of the inner tube finned. The rich solution flows inside the inner tube and vapours flow in the annular space between the tubes. The value of the overall heat transfer coefficient, referred to the outer surface area of the inner tube, is $U_{ve} = 30$ W/(m² K), and thus the exchanger surface area is $A_{ve} = Q'_{ve}/(U_{ve} \Delta T_{ln}) = 2.88$ m², where $\Delta T_{ln} = 13.35$ K.

Regenerative heat exchanger for the solutions. A concentric 'tube in tube' design was adopted. The value of the overall heat transfer coefficient is $U_{ve} = 600$ W/ (m² K), and thus the exchanger surface area is $A_{se} = Q'_{se}/(U_{se} \Delta T_{ln}) = 0.568$ m², where $\Delta T_{ln} = 26.87$ K.

Example 6.1.5

A diagram of the heat pump installation is shown in Figure 6.1.10, and its cycle in Figure 6.1.11. It was assumed that, due to the differences in saturation partial

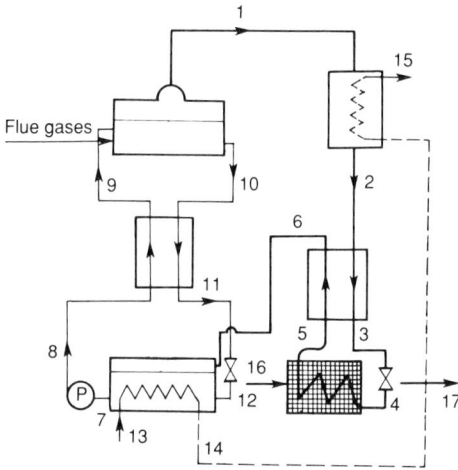

Figure 6.1.10   *Diagram of installation of an absorption cycle heat pump employing the R22–E181 pair of working fluids*

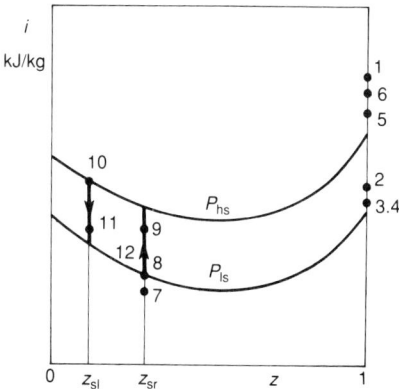

Figure 6.1.11   *Thermodynamic cycle of heat pump from Figure 6.1.10, in the i–z coordinates*

pressures of the working fluid and sorbent, desorption of the working fluid can be achieved without the rectification column.

The calculated extensive and intensive parameters are shown in Table 6.1.8, and the values of exchanged heat fluxes in Table 6.1.9. The power necessary to drive the solution pump is calculated as $N_p = 0.13 \, \text{kW}$. The COP_{aa} is:

$$COP_{aa} = (Q'_{ab} + Q'_{co})/(Q'_{de} + N_p) = (5.1 + 14.9)/(12.78 + 0.13) = 1.55$$

Table 6.1.8 *Extensive and intensive parameters calculated in Example 6.1.5 (point numbers correspond to points in Figure 6.1.10)*

Point number	Temperature T	Pressure (MPa)	Mass fraction	Mass flow (kg/s)	Enthalpy (kJ/kg)
1	348	2.00	1.000	0.030	440.00
2	328				270.00
3	305				240.00
4	260	0.34			As above
5	263				406.00
6	321		As above		426.00
7	320		0.275	0.320	487.97
8	329	2.00			488.27
9	393				659.55
10	433		0.200	0.290	724.87
11	338				535.90
12	330	0.34			As above
13	313	0.10	Water	0.238	167.52
14	325				220.85
15	333				251.27
16		0.10	Air	0.610	0.00
17	265				−8.10

Table 6.1.9 *Heat fluxes*

Part of installation	Heat flux (kW)	Comments
Absorber	5.10	High-temperature heat source
Condenser	14.90	High-temperature heat source
Evaporator	8.09	Low-temperature heat source
Desorber	12.78	External heat source (heat from burning oil)
Regenerative vapour heat exchanger	52.85	–
Regenerative solution heat exchanger	0.90	–

The calculations of exchanger surface areas were performed using methods and equations introduced in the preceding example. Values of the overall heat transfer coefficients and heat transfer areas are shown in Table 6.1.10.

Commentary. A larger value of the COP_{aa} in Example 6.1.5 results from the use of the R22–E181 pair and from the lack of exergy losses in the rectification column.

Table 6.1.10   *Surface areas of exchangers*

Part of installation	Overall heat transfer coefficient $(W/(m^2 K))$	Heat transfer area (m^2)
Absorber	800	3.140
Condenser	1000	0.680
Evaporator	30	19.830
Desorber[a]	2.4×10^4	0.127
Regenerative vapour heat exchanger	30	1.550
Regenerative solution heat exchanger	500	5.030

[a]Heat transfer coefficient (one side).

Example 6.2

This example illustrates a heat pump using water as the substance in the low-temperature source. The heat pump is used to heat a greenhouse providing 700 m² of arable area. To provide the initial design directives, the calculations were conducted for two different kinds of heat pumps: a mechanical vapour compression cycle heat pump using the R12 working fluid and a resorption cycle heat pump using the R717 (ammonia)–water pair.

Preliminary assumptions. The power in the high-temperature source $Q'_{hs} = 670 \, kW$ at temperature $T_{hs} = 333 \, K$, inlet water temperature $T_{water,inlet} = 313 \, K$, low-temperature source temperature $T_{ls} = 300 \, K$. Isentropic efficiency $\eta_{ic} = 0.75$, motor efficiency $\eta_{el} = 0.9$.

Example 6.2.1

The diagram of installation is shown in Figure 6.2.1, and its theoretical cycle in Figure 6.2.2. The following detailed assumptions were made: temperature difference at the cold end of the condensate subcooler 3 K, temperature difference at the evaporator outlet 5 K.

Cycle calculations

The calculations were performed graphically using the log *P–i* chart; the extensive and intensive parameters obtained are shown in Table 6.2.1, and values of the exchanged heat fluxes and the non-isentropic compression power in Table 6.2.2. The COP_{ac} is: $COP_{rc} = (Q'_{co} + Q'_{ev})/(N_{ni}/\eta_{el}) = 3.97$.

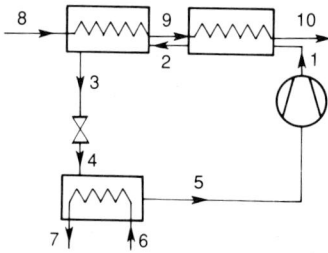

Figure 6.2.1   *Diagram of installation of a mechanical vapour compression cycle heat pump employing the R12 working fluid*

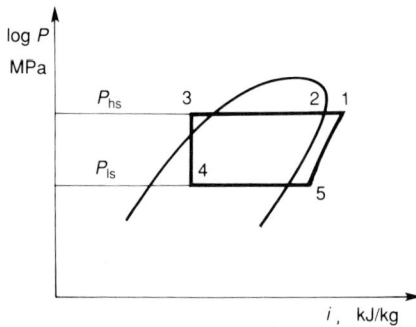

Figure 6.2.2   *Thermodynamic cycle of heat pump from Figure 6.2.1, in the log P–i coordinates*

Table 6.2.1   *Extensive and intensive parameters calculated in Example 6.2.1 (point numbers correspond to points in Figure 6.2.1)*

Point number	Temperature (K)	Pressure (MPa)	Mass flow (kg/s)	Enthalpy (kJ/kg)
1	361.1	1.69	4.16	698.00
1′	355.4	1.69	4.16	689.00
2	338.7	1.69	4.16	568.00
3	316.8	1.69	4.16	537.00
4	290.7	0.42	4.16	537.00
5	293.6	0.42	4.16	661.00
6	300.4	0.10	24.60	113.15
7	295.3	0.10	24.60	92.18
8	313.0	0.10	7.73	167.52
9	316.3	0.10	7.73	167.90
10	333.0	0.10	7.73	237.90

Table 6.2.2   *Heat fluxes and power supplied to the compressor*

Part of installation	Heat flux or non-isentropic compression power	Comments
Condenser	540.80	High-temperature heat source
Condensate cooler	129.20	High-temperature heat source
Evaporator	515.84	Low-temperature heat source
Compressor	153.92	Power delivered to compressor shaft

Exchanger calculations

Values of the overall heat transfer coefficients and surface areas of the exchangers are shown in Table 6.2.3.

Table 6.2.3   *Surface areas of exchangers*

Part of installation	Overall heat transfer coefficient $(W(m^2 K))$	Heat transfer area (m^2)
Condenser	1000	49.12
Condensate cooler	600	19.58
Evaporator	1400	54.48

Example 6.2.2

A diagram of the installation is shown in Figure 6.2.3, and its theoretical cycle in Figures 6.2.4 and 6.2.5. The following detailed assumptions were made: two-stage

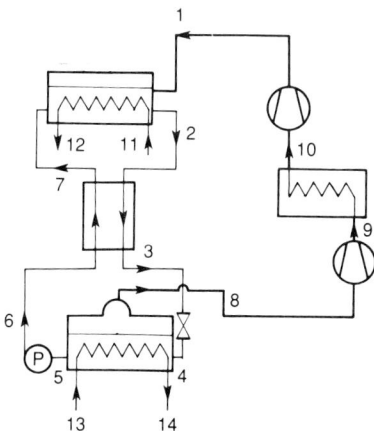

Figure 6.2.3   *Diagram of installation of a resorption cycle heat pump employing the R717–water pair of working fluids*

compression of the working fluid vapours with intercooling, isentropic efficiency η_{is} = 0.75 and mechanical efficiency of the solution pump $\eta_{m,sl} = 0.6$.

Cycle calculations

The calculations were performed graphically using the i–z and log P–i charts; results are shown in Table 6.2.4. Values of the exchanged heat fluxes and the compression power are shown in Table 6.2.5. The COP is:

$$COP_{rr} = Q'_{re}/((E'_{m1} + E'_{m2})/\eta_{el} + m_5 (i_6 - i_5)/(\eta_{m,sr}\eta_{el})) = 4.88$$

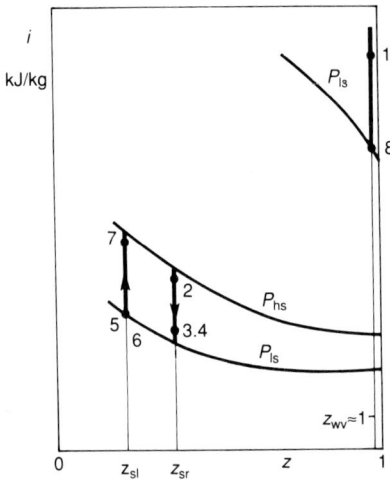

Figure 6.2.4   *Thermodynamic cycle of heat pump from Figure 6.2.3, in the i–z coordinates*

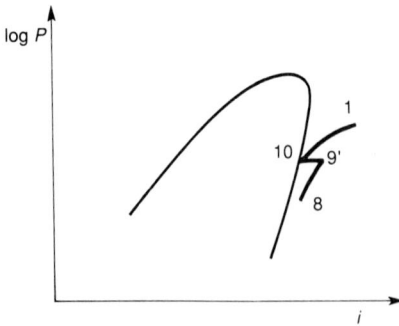

Figure 6.2.5   *Thermodynamic cycle of heat pump from Figure 6.2.3, in the log P–i coordinates*

Table 6.2.4   *Extensive and intensive parameters calculated in Example 6.2.2 (point numbers correspond to points in Figure 6.2.3)*

Point number	Temperature T	Pressure (MPa)	Mass fraction	Mass flow (kg/s)	Enthalpy (kJ/kg)
1	374	0.70	0.990	0.34	1977.0
1'	353	0.70	0.438	0.34	1940.0
2	336	0.70	0.438	2.08	155.0
3	316	0.70	0.438	2.08	7.0
4	292	0.12	0.438	2.08	7.0
5	293	0.70	0.330	1.74	4.0
6	293	0.70	0.330	1.74	6.0
7	338	0.70	0.330	1.74	184.0
8	299	0.12	0.990	0.34	1840.0
9	368	0.29	0.990	0.34	2040.0
9'	373	0.29	0.990	0.34	1990.0
10	294	0.29	0.990	0.34	1830.0
11	313	0.10	Water	8.07	168.0
12	333	0.10	Water	8.07	251.0
13	300	0.10	Water	29.43	113.0
14	295	0.10	Water	29.43	92.0

Table 6.2.5   *Heat fluxes and power supplied to compressors and mechanical pump*

Part of installation	Heat flux or electrical power (kW)	Comments
Resorber	670	High-temperature heat source
Desorber	618	Low-temperature heat source
Regenerative solution heat exchanger	310	–
Interstage cooler	71	–
First-stage compressor	69	Power delivered to compressor shaft
Second-stage compressor	51	Power delivered to compressor shaft
Lean solution pump	3	Power delivered to pump shaft

Exchanger calculations

Values of the overall heat transfer coefficients and surface areas of the exchangers are shown in Table 6.2.6.

Table 6.2.6   *Surface areas of exchangers*

Part of installation	Overall heat transfer coefficient (W/(m² K))	Heat transfer area (m²)
Resorber	800	70.35
Desorber	1500	100.00
Regenerative solution heat exchanger	600	42.42

Commentary. Because in a resorption cycle heat pump the changes of phases are conducted at constant pressures, not at constant temperatures, this leads to lower exergy losses.

Example 6.3

Ejector-based heat pumps have not been given much attention in this book, as they cannot be profitable. The topic is limited to an example, which explains the principle of operation of the ejector, but without detailed calculations. These can be found in specialist works, e.g. [120], where extensive material on ejector calculations can be found, and [121], where applications of ejectors in refrigeration cycles are discussed.

In this example we utilize the optimization calculations of a theoretical cycle [21], in which throttling losses in the control valve and ejector losses are taken into account; other losses are neglected. The working fluid is the R11, temperature range: $T_{ls} = 283.2$ K, $T_{hs} = 316.5$ K, $T_{ext} = 366.6$ K.

An ideal ejector implements both the isentropic expansion of the driving fluid and isentropic compression of the fluid being driven. Since in the case under consideration both the fluids are identical chemically, the processes taking place in the ejector can be illustrated on the *i–s* chart (Figure 6.3.1a). After mixing the stream m_1 at state 7 and stream m_2 at state 1, one arrives at a state 2 located on the

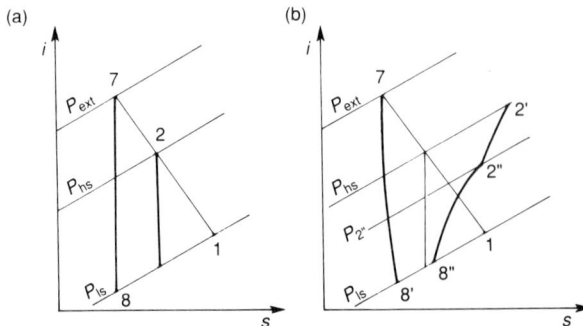

Figure 6.3.1   *Thermodynamic processes in an ejector:* (a) *ideal;* (b) *real*

mixing line 721. According to the principles of conservation of mass, energy and entropy (without its production), the location of point 2 is defined by a relationship $m_1/(m_1 + m_2)$, or the ratio of distance 72 to distance 71. Assuming that the ejector operates without losses, it is easy to define the mass flow m_1, for the same magnitude of heat flux delivered to the high-temperature source of the heat pump, as in the case of a cycle with losses.

In a real ejector, irreversible processes and the accompanying losses take place, and the design calculations of an ejector are basically reduced to their description. The irreversible processes taking place in an ejector are: expansion in the nozzle, mixing in the mixing chamber (a cylindrical section of the ejector) and compression in the diffuser; these processes are illustrated in Figure 6.3.1b. Irreversibility in the expansion nozzle can be expressed in terms of efficiency, defined by the relationship $\eta_{m,en} = (i_7 - i_8')/(i_7 - i_8)$. Its value is not much different from unity (following the quoted reference it was taken here as $\eta_{m,en} = 0.97$); it has, nevertheless, a significant effect on the exergetic efficiency of a heat pump. Efficiency in the diffuser is defined by the relationship $\eta_{m,di} = (i_{2'''} - i_2)/(i_{2'} - i_{2''})$; it was assumed here, after the quoted reference, to be $\eta_{m,di} = 0.7$.

The calculations require great accuracy, as during the cycle calculations, particularly when calculating the losses, values of thermodynamic parameters are often decided by the third decimal places in the numbers involved. As is known, these values differ between various sources; in view of the lack of information on the sources in the quoted reference, data from [122] were used here.

A diagram of the heat pump installations is shown in Figure 6.3.2, and the approximate cycle in Figure 6.3.3. The results of calculations can be assembled into four groups: extensive parameters of the cycle, characteristic data of the ejector,

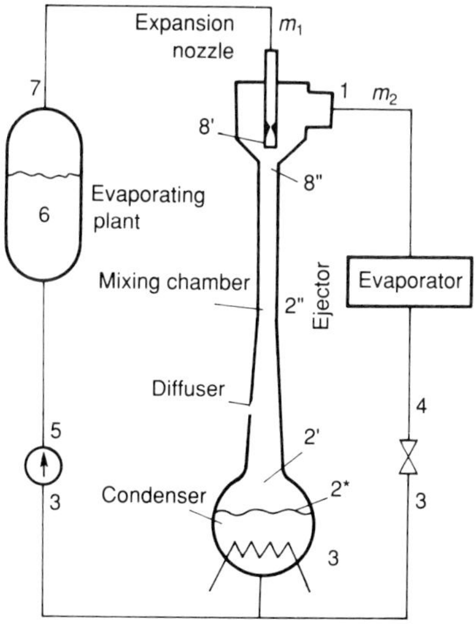

Figure 6.3.2 *Diagram of an ejector-based heat pump*

intensive parameters, and losses in irreversible processes (Tables 6.3.1, 6.3.2, 6.3.3 and 6.3.4, respectively).

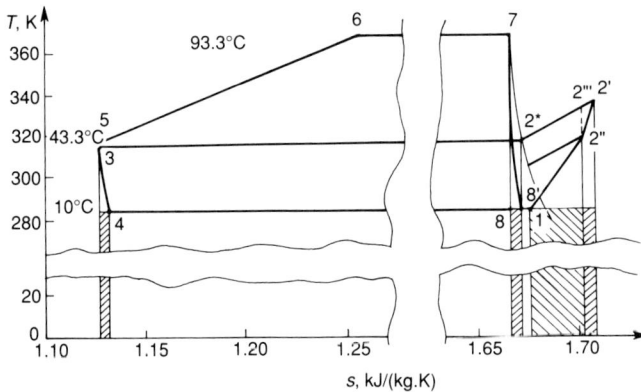

Figure 6.3.3   *Approximate cycle of an ejector compression cycle heat pump*

Table 6.3.1   *Extensive cycle parameters*

Thermal power:
  at the high-temperature heat source level of the heat pump:
  $Q'_{hs} = (m_1 + m_2)(i_{2'} - i_3) = (327.5 + 124.5)(724.5 - 539.2) = 83.76$ MW
  supplied to the evaporating plant:
  $Q'_{ext} = m_1(i_7 - i_5) = 327.5(737.0 - 539.5) = 64.25$ MW

Mass flows of working fluids:
  in the heat engine cycle $m_1 = 327.5$ kg/s
  in the actual heat engine cycle $m_2 = 124.5$ kg/s
  in the condenser $(m_1 + m_2) = 452.0$ kg/s

Primary energy ratio $Q'_{hs}/Q'_{ext} = 1.3037$

Table 6.3.2   *Characteristic data of the ejector*

Diameters:
  throat diameter in expansion nozzle   0.344 m
  cylindrical diameter of mixing chamber   0.770 m

Velocities:
  at exit from expansion nozzle   291.9 m/s
  at exit from cylindrical part of the mixing chamber   131.4 m/s

Compression ratio: $P_2^*/P_1 = 3.18$

Ejection ratio: $m_2/m_1 = 0.38$

Table 6.3.3   *Intensive parameters of the cycle*

Point number	T (K)	P (MPa)	i (kJ/kg)	s (kJ/(kg K))
1	283.0	0.06127	696.9	1.684
2′	337.0		724.5	1.711
2″	320.0	0.13640	717.0	1.703
2*			713.0	1.673
3	316.5	0.19288	540.6	1.128
4			540.6	1.132
5			539.5	1.128
6			583.1	1.265
7	366.5	0.71582	736.0	1.668
8			692.6	1.668
8′			693.3	1.673
8″			694.3	1.674

Table 6.3.4   *Listing of losses*

Type of loss	Formula	MW	%
Losses in control valve	$\delta B'_{cv} = (s_4 - s_3)m_1 T_{ls}$	0.141	2.7
Losses in ejector			
in the expansion nozzle	$\delta B'_{en} = (s_8, - s_7)m_1 T_{ls}$	0.464	9.2
in the mixing chamber	$\delta B'_{mc} = (s_2,(m_1 + m_2) - (s_1 m_2 + m_1 s_8,))T_{ls}$	3.366	68.0
in the diffuser	$\varrho B'_{di} = (s_2, - s_2)(m_1 + m_2)T_{ls}$	1.024	20.0
Total		4.994	100
Energy delivered at the high-temperature heat source		8.950	
Driving exergy supplied to the evaporation plant		13.820	
		4.92	

For comparison, results of calculations of an ideal reverse Rankine cycle and a theoretical cycle without losses in the ejector are shown below.

Ideal reverse Rankine cycle:

$$COP_{iR} = ((T_{ex} - T_{ls})/T_{ext}) \, T_{hs}/(T_{hs} - T_{ls})$$
$$= ((366.5 - 283.0)/366.5) \, 316.5/(316.5 - 283.0) = 21.5$$

Theoretical cycle (Figure 6.3.4):

Assuming that $i_2 = i_{2^*}$, heat flux released at the high-temperature level is:

$$(m_{1t} + m_{2t}) (i_{2^*} - i_3) = Q'_{hs}$$

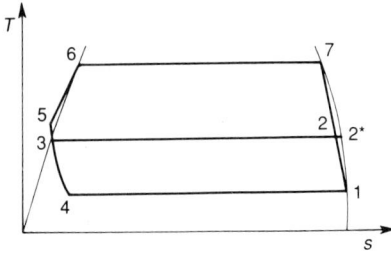

Figure 6.3.4   *Theoretical cycle of an ejector compression heat pump*

energy balance of the ejector can be written as:

$$(m_{1t} i_7 + m_{2t} i_1) = (m_{1t} + m_{2t})i_{2*}$$

Putting into the above equations values of i_1, i_{2*}, i_3, i_7, Q'_{hs} (Tables 6.3.1 and 6.3.2), mass flow rates are:

$$m_{1t} = 200.11 \, \text{kg/s and } m_{2t} = 285.73 \, \text{kg/s}$$

External heat flux is defined as follows:

$$Q'_{ext} = m_{1t}(i_7 - i_5)$$

For values of i_5 and i_7 given in Table 6.3.3:

$$Q'_{ext} = 200.11 \, (736 - 539.5) = 39.32 \, \text{MW}$$

and PER

$$\text{PER} = Q'_{hs}/Q'_{ext} = 83.76/42.89 = 1.949$$

Example 6.4

Waste heat from the condenser (dephlegmator) of a rectification column can be used as the low-temperature heat source. Three variants of such designs follow.

Example 6.4.1

This example illustrates an absorption cycle heat pump, working on the R718–LiBr (water–lithium bromide) pair, attached to a isopropyl alcohol rectification column. A diagram of the installation is shown in Figure 6.4.1. The characteristic parameters of the heat pump resulting from such an arrangement are as follows: low-temperature source temperature $T_{ls} = 345 \, \text{K}$, high-temperature source $T_{hs} = 390 \, \text{K}$, thermal power at the high-temperature source level $Q'_{hs} = 600 \, \text{kW}$. The thermodynamic cycle is illustrated in the log P–$1/t$ plane (Figure 6.4.2). Such an installation was built by the Hitachi Zosen Company. In terms of mechanical

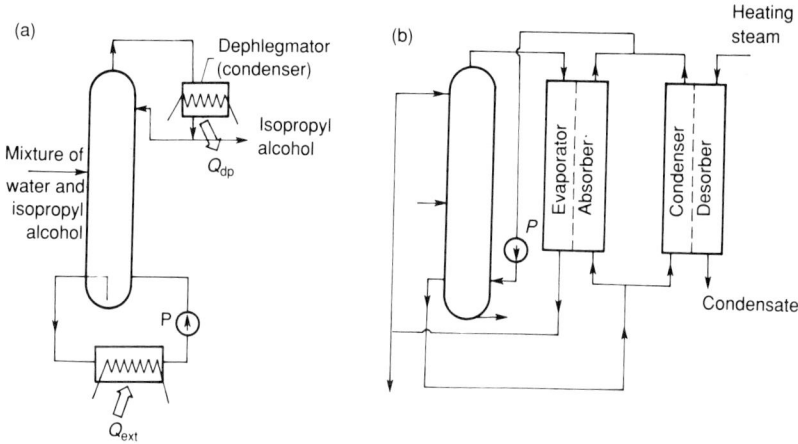

Figure 6.4.1   *Diagrams of the rectification column and heat pump installations:* (a) *column without a heat pump installed;* (b) *column with heat pump attached*

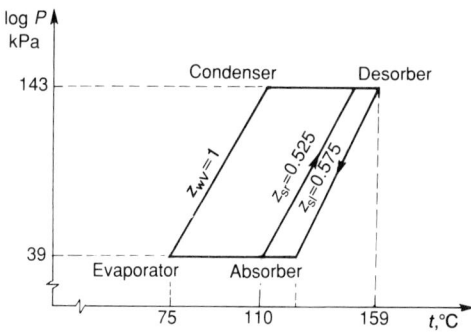

Figure 6.4.2   *Thermodynamic cycle of heat pump of Figure 6.4.1, in the log P–1/T coordinates*

design the heat pump consists of two vertical cylindrical tanks. The first one contains the evaporator and absorber; the second contains the condenser and desorber. The lithium bromide solution is doped with octyl alcohol ($C_8H_{17}OH$) to improve the surface wettability and with lithium chromate (Li_2CrO_4) to reduce the corrosiveness of the water solution of lithium bromide. It can be added here that by installing the heat pump in this application the demand for steam from the boiler was reduced by 42% (Figure 6.4.3).

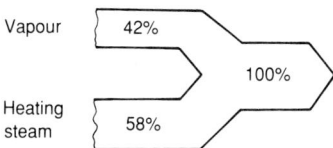

Figure 6.4.3   *Flow diagram of energy for installation of Figure 6.4.1*

Example 6.4.2

In this example we discuss the use of a heat transformer based on the pair R718–lithium bromide. Data are taken from [69]. The external source is provided by steam leaving the column. Its temperature is $T_{ext} = 371$ K, and the energy flux collected from this steam is $Q_{ext} = 5$ MW. The high-temperature source is characterized by temperature $T_{hs} = 409$ K and heat flux $Q'_{hs} = 2.35$ MW. The temperature of the water cooling the condenser is $T_{ls} = 298$ K.

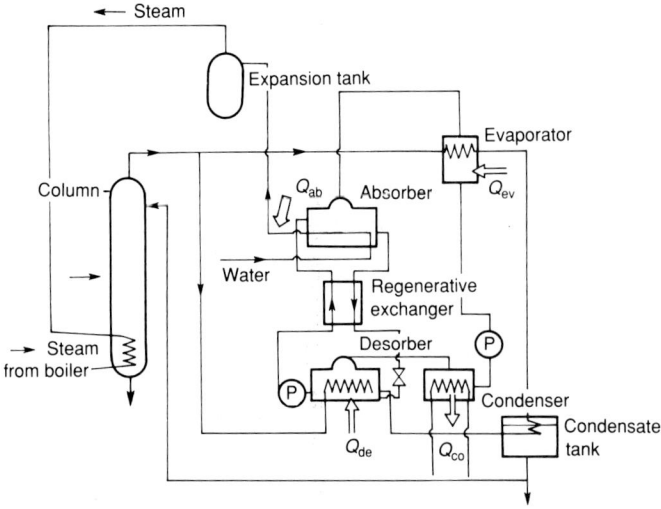

Figure 6.4.4   *Diagram of installation of a rectification column and a heat transformer*

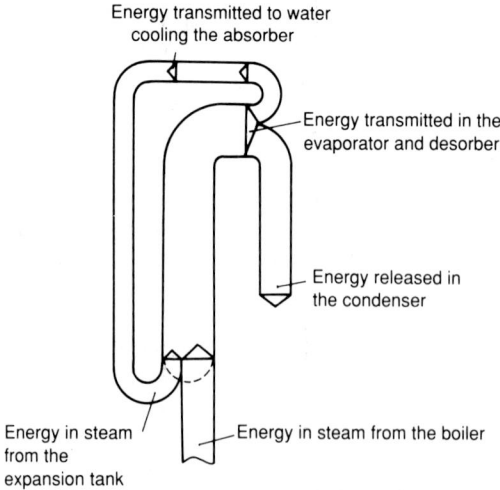

Figure 6.4.5   *Flow diagram of energy for the installation of Figure 6.4.4*

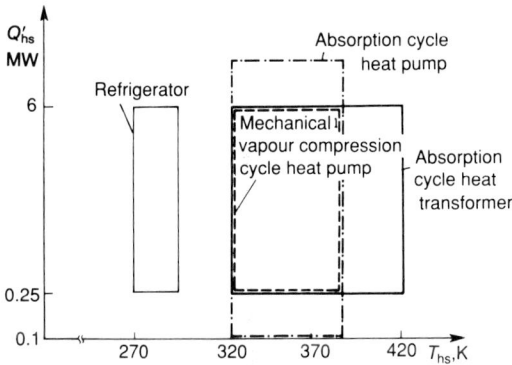

Figure 6.4.6 *Application ranges of Hitachi Zosen heat pumps*

Table 6.4.1 *General characteristics of absorption cycle heat pump and heat transformer employing the R718–lithium bromide pair as working fluid*

Heat source	Absorption cycle heat pump	Heat transformer
External	High temperature $T_{ext} = T_{de} > 420$ K	Waste heat $T_{ext} = T_{de} = T_{ev}$ $T_{ext} \in (350–380)$ K
High temperature	$T_{hs} = T_{ab} = T_{co}$ $T_{hs} \in (330–390)$ K	$T_{hs} = T_{ab}$ $T_{hs} \in (350–420)$ K
Low temperature	$T_{ls} = T_{ev}$ $T_{ls} \in (300–320)$ K	$T_{ls} = T_{co}$ $T_{ls} \in (280–290)$ K Substance – water
COP	$COP_{ra} = 1.8$	$COP_{rt} = 0.45$
Application	Recovery of low-temperature heat from artificial or natural sources	Recovery of medium-temperature waste heat

A diagram of the installation, together with the rectification column, is shown in Figure 6.4.4. Heat is supplied to the desorber and evaporator. Water leaving the absorber is directed to an expansion tank, and the steam generated in this tank is used to heat the rectification column.

Figure 6.4.5 shows the flow diagram for energy. It should be stressed here that the capital cost of the installation was recovered within a few months.

A comparison of parameters of the absorption cycle heat pump and the heat transformer is shown in Figure 6.4.6 and in Table 6.4.1. One should mention here that by using the heat transformer, it is possible to obtain temperatures above 400 K in the high-temperature source. This necessitates, however, the use of cold water $T_{en} \in (285–290)$ K. Possible corrosion inhibitors are discussed in [26].

Example 6.4.3

Examples 6.6.1 and 6.6.2 illustrate the application of a screw compressor Mycom STM to compress water vapour. This compressor cannot, however, be used to compress other vapours, e.g. ethanol vapour.

In order to recover heat from the rectification column, a water (R718) based heat pump, e.g. a mechanical vapour compression cycle heat pump, needs to be used. In contrast to the direct vapour compression, one must use a heat exchanger, in which on one side the ethanol vapours are being condensed, while on the other side the working fluid is evaporated. Thus additional exergy losses arise, which are not taken into account at this stage of the calculations. Water is a very advantageous working fluid because of the range of temperatures in this installation. The heat pump installation is nearly standard, the novelty is an additional internal water flow to cool the compressor. The diagram of heat pump installation, together with the rectification column, is shown in Figure 6.4.7. Without going into details of the rectification process, which takes place at a pressure close to atmospheric, the temperature in the condenser is 348.1 K, the temperature in the evaporator is 385.2 K, and the energy flux delivered in the evaporator is 1965 kW. The compressor is cooled with water at state 3 (Figure 6.4.8) and mass flow rate 0.09 kg/s. This water is injected to a zone of already elevated pressure in the compressor, so that the generated vapours need only be compressed through part of the pressure range. A simplifying assumption is made that the demand for compression work is the same as for the compression through the total pressure

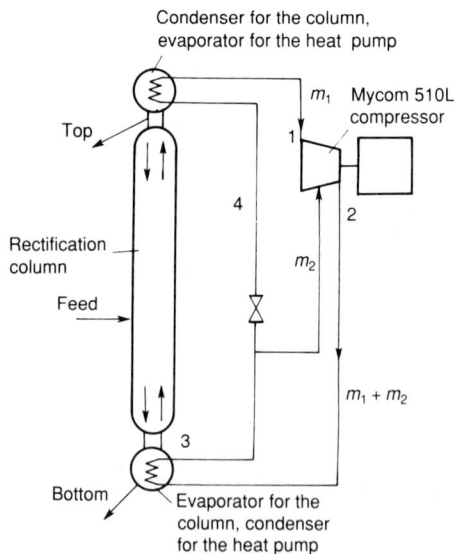

Figure 6.4.7 *Diagram of an installation for heat recovery from the dephlegmator (condenser) of a rectification column by means of a mechanical vapour compression cycle heat pump employing the R718 working fluid*

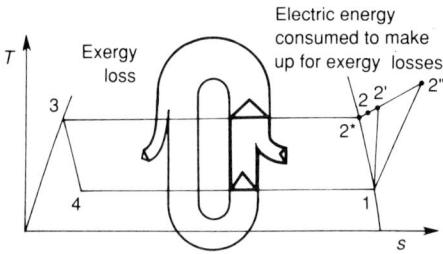

Figure 6.4.8 *Heat pump cycle, with a schematic representation of transferring heat from the high-temperature heat source to the low-temperature heat source*

difference of 1/3 of the mass flow of vapour generated by evaporation of the cooling water. Also a value of the isentropic compression efficiency was assumed: $(i_{2'} - i_1)/(i_{2''} - i_1) = 0.7$. The final state 2 of the vapour leaving the compressor (Figure 6.4.8) is derived from a simplified condition, stating that the total heat flux is recovered at constant pressure outside the compressor: $m_2(i_2 - i_3) = (m_1 + m_2)(i_{2''} - i_2)$. The values of extensive and intensive parameters are listed in Tables 6.4.2 and 6.4.3.

Table 6.4.2 *Extensive parameters of the cycle*

Mass flows:
  working fluid $m_1 = 0.755$ kg/s
  compressor cooling water $m_2 = 0.09$ kg/s

Thermal power at the high-temperature heat source level of the heat pump (in the evaporator of the rectification column):
  $Q'_{hs} = (m_1 + m_2)(i_2 - i_3) = (0.755 + 0.09)(2796.0 - 469.8) = 1965$ MW

Power in the low-temperature heat source (in the condenser of the rectification column:
  $Q'_{ls} = m_1(i_1 - i_4) = 0.755(2634.4 - 469.8) = 1635$ MW

Drive power of the cycle:
  $N = (m_1 + (1/3)m_2)(i_{2'} - i_1) = 0.785(3043.1 - 2634.8) = 320.5$ kW

Losses:
  In the compressor (lost work):
  $(1 - 0.7)(i_{2'} - i_1)(m_1 + (1/3)m_2) = 0.3(2920.6 - 2634.8)0.785 = 67.30$ kW
  in the valve:
  $(m_1 + m_2)(s_3 - s_4)T_4 = 0.845(1.462 - 1.44)348.1 = 6.47$ kW

Coefficient of performance of the cycle:
  $COP = Q'_{hs}/N = 1943/320.5 = 6.06$

Table 6.4.3   *Intensive parameters of the cycle*

Point number	T (K)	P (MPa)	i (kJ/kg)	s (kJ/(kg K))
1	348.1	0.03835	2634.8	7.6820
2'	503.2	0.03835	2920.6	7.6820
2''	556.1	0.03835	3043.1	7.9390
2	445.1	0.15250	2796.0	7.3922
3	385.1	0.15250	469.8	1.4400
4	348.1	0.03835	469.8	1.4620

Example 6.5

This example illustrates heat recovery from the condenser of a mechanical vapour compression refrigeration plant used in freeze-drying of meat. The example is based on data presented in [78]. A mechanical vapour compression cycle heat pump is proposed, with R12 as a working fluid; the temperature in the heat pump condenser $T_{hs} = 333.0$ K, and the temperature in the refrigerator condenser $T_{ls} = 268.0$ K. A diagram of the installation is presented in Figure 6.5.1. Freeze-drying relies on the sublimation of moisture contained in the material (meat). It is placed in a heated sublimation chamber, which is under vacuum as a result of being connected to a vacuum pump through a refrigeration chamber (Figure 6.5.1). In the refrigeration chamber the generated vapours are desublimed

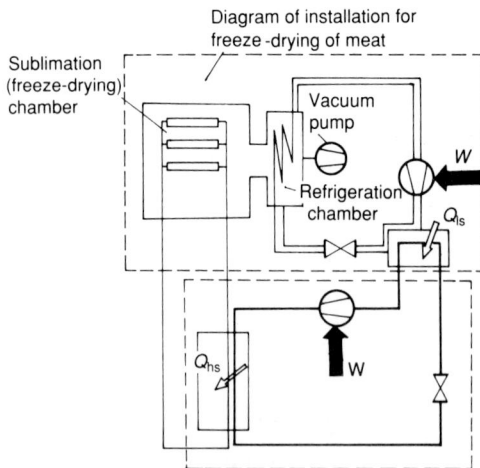

Figure 6.5.1   *Diagram of installation for freeze–drying of meat and a cooperating mechanical vapour compression cycle heat pump employing the R12 as working fluid*

(freeze-condensed out). Heat transferred and the driving work are referred to 1 kg of the working fluid in the heat pump cycle. The heat pump releases in its condenser $q_{co} = 130$ kJ/kg, and the supplied work is $w_r = 48$ kJ/kg; thus the $COP_{rc} = 2.7$. Application of this heat pump makes it possible to reduce the energy demand for freeze-drying by a quarter of the original value.

Example 6.6

An example of the application of mechanical compressors for vapour compression, presented in [94], proposes the use of screw compressors manufactured by the Mayehawa Company, which in 1983 introduced the Mycom STM range of compressors for water vapour (steam). Compressors in this range can compress water vapour from the temperature range 60–120°C up to the range of 120–180 (200)°C, with a compression ratio of 3.8, volumetric flow rates at the inlet of 1–10 m^3/s, and at 3000–7000 rev/min.

Example 6.6.1

This example illustrates the compression of vapours originating from the mash vat during brewing beer. The compressed vapours are directed to a heat exchanger (Figure 6.6.1), where they sustain the boiling process without any heat being delivered from outside. The open cycle of the heat pump is implemented in the compressor and heat exchanger, which on the one hand is a vapour condenser, and on the other hand a mash evaporator. In this exchanger heat is transferred from the

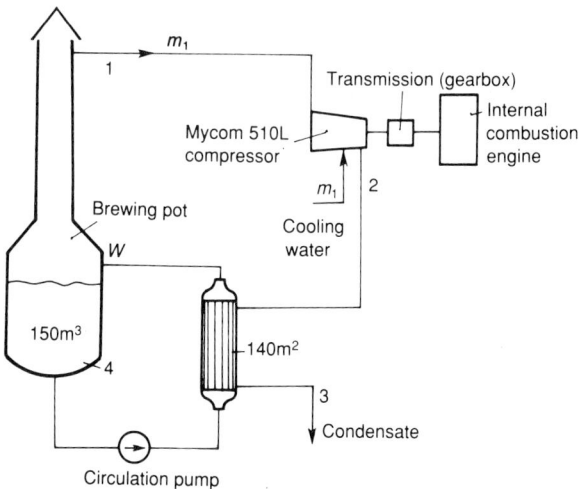

Figure 6.6.1 *Diagram of installation for mechanical compression of vapours, cooperating with a mash brewing pot*

high-temperature source to the low-temperature source, as is shown in Figure 6.6.2, which illustrates the open cycle. The extensive and intensive parameters of the cycle are given in Tables 6.6.1. and 6.6.2.

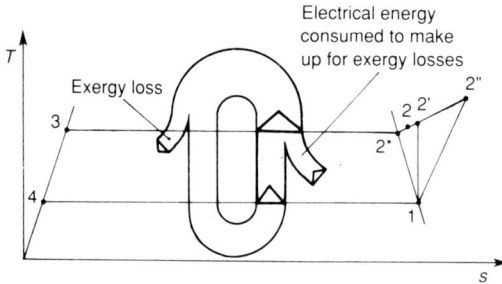

Figure 6.6.2   *Open cycle of the heat pump*

Table 6.6.1   *Extensive parameters*

Mass flows:
   vapour removed $m_1 = 2.78$ kg/s
   water cooling the compressor $m_2 = 0.22$ kg/s

Thermal power at the high-temperature heat source level of the heat pump:
   $Q'_{hs} = (m_1 + m_2)(i_2 - i_3) = (2.78 + 0.22)(2728.5 - 571.9) = 6.47$ MW

Power necessary to drive the cycle:
   $N = (m_1 + (1/3)m_2)(i_{2''} - i_1) = (2.78 + 0.07)(3000.4 - 2675.8)$
   $\qquad\qquad\qquad\qquad\qquad\qquad = 28.5 \times 324.6 = 0.925$ MW

Shaft power of the internal combustion engine:
   $N_{Ht} = N/\eta_m = 0.925/0.92 = 1.00$ MW

Coefficients of performance of the cycle:
   theoretical: $COP_{ac} = Q'_{hs}/N = 6.47/0.925 = 6.99$
   realized: $COP_{rc} = Q'_{hs}/N_{Ht} = 6.47/1.00 = 6.47$

Compared with the primary energy input, assuming the efficiency of the internal combustion engine to be 0.35 (which is a large value):
   $PER = Q'_{hs}/(N_{Ht}/0.35) = 2.26$

Example 6.6.2

This illustrates the application of vapour compression in a multistage evaporation installation used to concentrate sugar solution. A diagram of the installation is shown in Figure 6.6.3. Both the installation discussed here, and the classic installation employ a well-developed system to regenerate the heat from hot solution and from the hot vapours; this heat is used to heat the feedstock. Mass flows of the solution and condensate are significant, and the regenerated heat fluxes are correspondingly high. These fluxes determine the mass and energy flows

Table 6.6.2 *Intensive parameters*

Point number	T (K)	P (MPa)	i (kJ/kg)	s (kJ/(kg K))
1	373.2	0.1013	2675.8	7.355
2'	492.7	0.3222	2903.0	7.355
2"	548.4	0.3222	3000.4	7.575
2*	409.2	0.3222	2728.5	6.969
2	435.2	0.3222	2833.4	7.091
3	409.2	0.3222	571.9	1.697
4	373.2	0.1013	419.0	1.307

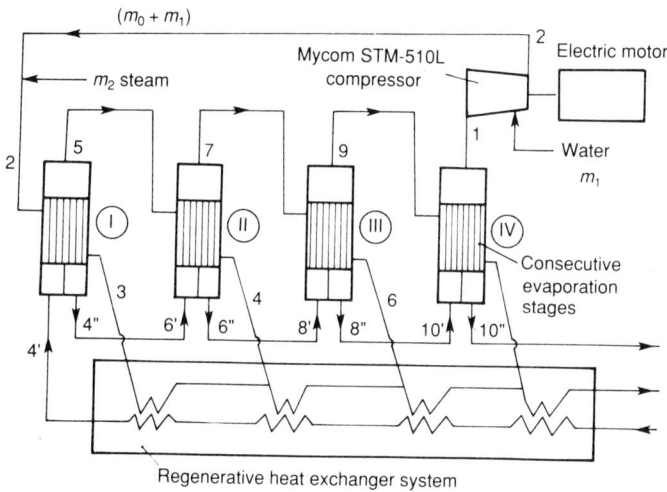

Figure 6.6.3 *Diagram of installation for mechanical compression of vapours, cooperating with a multistage evaporation plant*

in a multistage evaporation installation. In this discussion, however, we do not have to consider the details of the evaporation installation, including the issues of its multistage nature, but will isolate the heat pump itself from the installation. The heat pump consists of a compressor, first stage evaporator, which acts as a condenser in the heat pump cycle, and the fourth-stage evaporator, which acts as the heat pump evaporator.

The heat pump cycle is shown in Figure 6.6.4, where heat transmission from the high-temperature source to the low-temperature source is also indicated schematically. The laws of conservation of mass (of water and sugar) and of energy must apply in the whole installation. The mass balance of water and the energy balance are closed by an additional supply of steam at a rate m_2 and with enthalpy i_2. The intensive and extensive parameters of the open cycle of such heat pumps are shown in Tables 6.6.3 and 6.6.4.

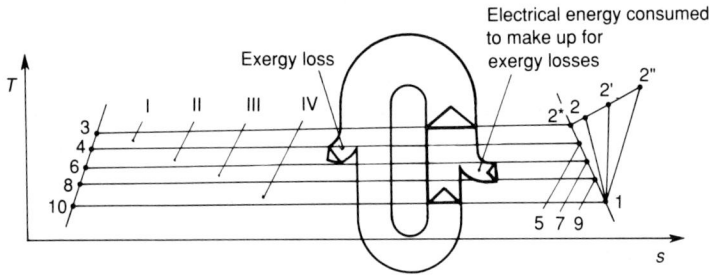

Figure 6.6.4   *The open heat pump cycle showing the temperature at individual stages of the evaporation plant, with a schematic representation of transferring heat from the high-temperature heat source to the low-temperature heat source*

Table 6.6.3   *Extensive parameters*

Mass flows:
    vapour collected from the first stage
        $m_0 = 2.594$ kg/s
    water cooling the compressor:
        $m_1 = 0.196$ kg/s
    additional steam having parameters as at point 2*:
        $m_2 = 0.104$ kg/s (calculated from heat losses)
    feedstock, 18%; 30°C; 13.93 kg/s
    product, 66.6%; 100°C; 3.77 kg/s
    condensate, 35°C; 10.46 kg/s

Thermal power at the high-temperature heat source level:
    $Q'_{hs} (m_0 + m_1)(i_2 - i_3) = 2.79(2741.4 - 558.9) = 6.09$ MW

Power necessary to drive the cycle:
    $N = (m_0 + (1/3)m_1)(i_{2''} - i_1) = (2.594 + 0.059) (2973.4 - 2675.8) = 789.53$ kW

Coefficient of performance of the cycle:
    $Q'_{hs}/N = 6090/789.53 = 7.71$

Table 6.6.4   *Intensive parameters*

Point number	T (K)	P (MPa)	i (kJ/kg)	s (kJ/(kg K))
1	373.2	0.1013	2675.8	7.3550
2'	482.7	0.2953	2884.1	7.3550
2''	526.9	0.2953	2973.4	7.5434
2*	406.2	0.2953	2724.4	6.9970
2	413.0	0.2953	2741.4	7.0320
3	406.2	0.2953	558.9	1.6659
10	399.2	0.2393	529.2	1.5920

Example 6.7

A mechanical vapour compression cycle heat pump is used to utilize waste heat from sewage originating from a dairy bottle washing plant (30 000 bottles per hour). The low-temperature source is provided by the sewage water having inlet temperature $T_{ls,inlet} = 311$ K and supplying a heat flux $Q'_{ls} = 170$ kW, and the substance of the high-temperature source is water at a maximum temperature $T_{hs,max} = 353$ K and heat flux $Q'_{hs} = 240$ kW. Electrical power supplied to the compressor motor is $N_{el} = 70$ kW.

A diagram of the bottle washing plant and of the heat pump installation is shown in Figure 6.7.1. The bottle washing procedure is as follows. Bottles are placed

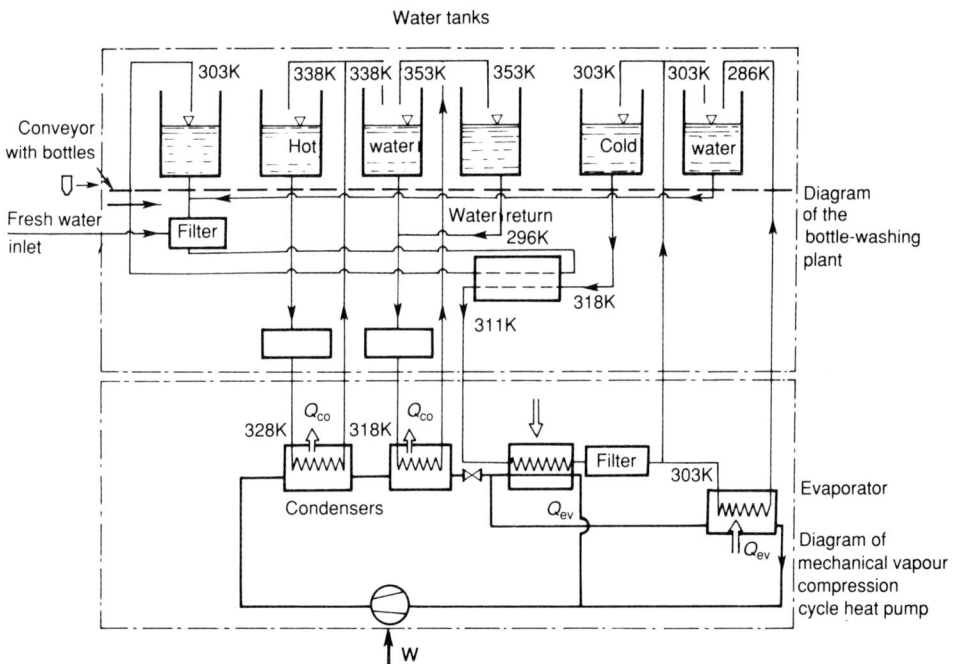

Figure 6.7.1 *Diagram of installation of a mechanical vapour compression heat pump and a bottle-washing plant*

upside down on a conveyor and passed underneath the retention tank containing washing water. The residence time of the bottles in the washer is 3–4 min. To meet the necessary hygienic demands, bottles are heated during washing to a temperature between 343 and 353 K. At the entry point their temperature is equal to that of the environment and at the exit, due to the required conditions of filling them with milk, their temperature must be 283 K. This results in a need for variations in the washing water temperature; these changes are made stepwise.

The use of a heat pump makes it possible to obtain both the hot and cold water.

To reduce losses resulting from temperature differences, a two-component working fluid was employed, consisting of a blend of the R12 and R22 working fluids.

Figure 6.7.2 shows a flow diagram of energy for the discussed installation. It was possible to reduce demand not only for energy, but also for mains water. Such a heat pump was installed in 1978 in England; capital costs were recovered within approximately two years.

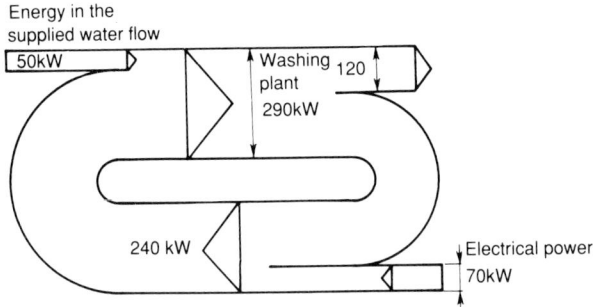

Figure 6.7.2   *Flow diagram of energy for installation of Figure 6.7.1*

Example 6.8

Application of a heat pump co-operating with a central heating installation enables the utilization of waste heat for district heating of large urban agglomerations. The high-temperature source of the heat pump is provided by the return water in a central heating installation. Examples of such applications are illustrated by three solutions implemented several years ago in Sweden.

Example 6.8.1

The heat recovery from technological steam used in a sugar refinery, with a daily yield of 600 tons of sugar, was realized by employing a mechanical vapour compression cycle heat pump with R12 as the working fluid, and with a through-flow compressor driven by a steam turbine. The low-temperature source is provided by technological steam generated in the initial stage of sugar beet processing. Its temperature is $T_{ls} = 338\,K$ and the energy flux provided by this steam is $Q'_{ls} = 10.5\,MW$. The high-temperature source is provided by the heat pump and turbine condensers connected in series. The temperature of the water at the inlet to the heat pump condenser is $T_{water,inlet} = 333\,K$, and the exit temperature is $T_{water,outlet} = 353\,K$. This water is next heated in the turbine condenser by a further $\delta T = 7\,K$. The heat flux is transferred to water circulating in a district central heating installation of an estate with 6000 residents. The power of the turbine driving the compressor is 2.1 MW.

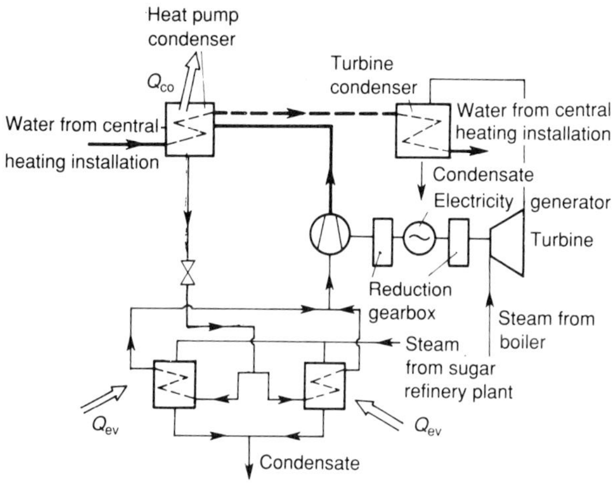

Figure 6.8.1 *Mechanical vapour compression cycle heat pump cooperating with a central heating installation (low-temperature heat source is provided by technological steam from sugar refinery)*

A diagram of the installation is shown in Figure 6.8.1. Because of the high power involved, two evaporators are used in series. An electricity generator of 0.5 MW power was also installed to serve the local demand. At rated conditions the heat flux transferred in the high-temperature source is $Q'_{hs} = 16.9$ MW and the required heat flux contained in steam driving the turbine (steam pressure $P = 2.8$ MPa and $T = 650$ K) is $Q'_{ext} = 6.4$ MW; thus PER = 2.64. When the demand for electric energy $N_{el} = 0.5$ MW, then $Q'_{hs} = 18.9$ MW and PER = 2.23.

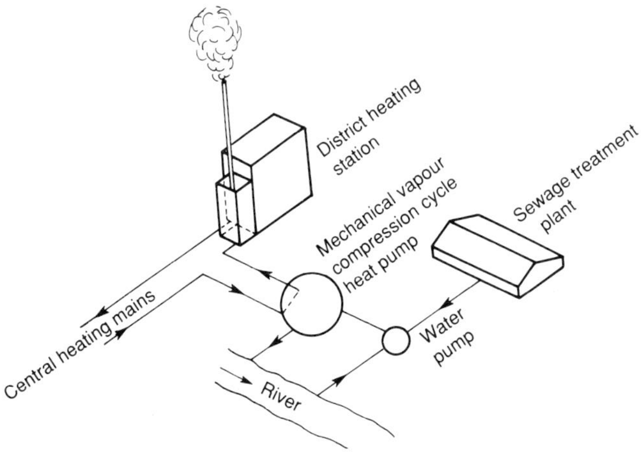

Figure 6.8.2 *Mechanical vapour compression heat pump cooperating with central heating installation (low-temperature heat source is provided by sewage supplemented by river water)*

Example 6.8.2

The waste heat source is provided by purified waste water supplemented, if necessary, by water from the river. Temperatures of the low- and high-temperature sources vary throughout the year, and so: $T_{ls} = 281–288\,K$ and $T_{hs} = 321–333\,K$, respectively. The heat fluxes, averaged annually, are correspondingly $Q'_{ls} = 2.1\,MW$ and $Q'_{hs} = 3.3\,MW$. A through-flow compressor was used to compress the working fluid vapours.

A diagram of the installation is shown in Figure 6.8.2. Because of the corrosive nature of the low-temperature heat source substance, an external cascade flow of waste water over evaporator tubes was used, with horizontal tubes made from galvanized mild steel. The capital cost of installation (in 1981 prices) was 770 000 US dollars and was recovered within three years.

Example 6.8.3

The source of waste heat is provided by water cooling the flue gases leaving a solid waste incinerator with a daily throughput of 750 tonnes of waste. The temperature of water is within the range $T_{ls} = 298–308\,K$. This water provides the low-temperature heat source for an absorption cycle heat pump. Heat flux $Q'_{hs} = 50–53$ (MW) is transferred to the return water, at a temperature $T_{ls} = 333\,K$,

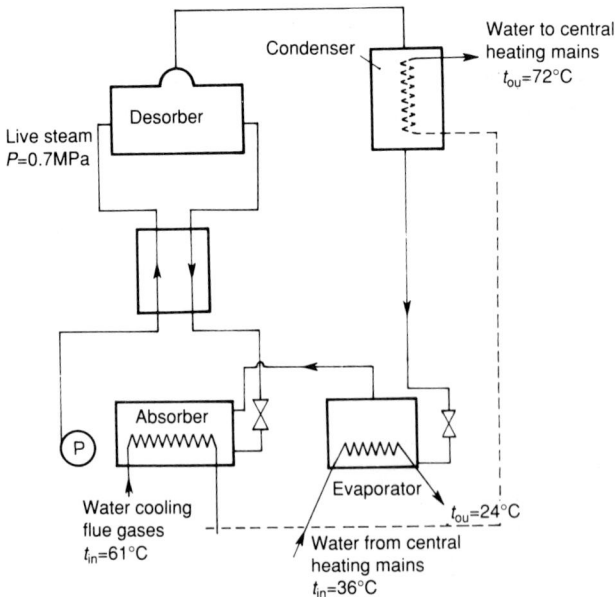

Figure 6.8.3 *Substances of heat sources in a large power absorption cycle heat pump*

in a central heating installation. The increase in temperature of this water is $\delta T_{hs} = 15$ K. The COP of this heat pump is $COP_{ra} = 1.61$.

Live heating steam (at a pressure 0.7 MPa), also produced in the incinerator installation, is delivered to the desorber. It is worth mentioning that in Uppsala (Sweden), where this heat pump is installed, 40% of thermal energy demand is generated from waste incineration.

A diagram of the installation is shown in Figure 6.8.3. The R718–lithium bromide pair is used as the working fluid. To minimize the exergy losses resulting from temperature differences, the installation is built as two heat pumps connected in series.

Example 6.9

A network of heat exchangers, designed using the Pinch Technology method [124], is shown in Figure 6.9.1a. Figure 6.9.1c shows the temperature changes in temperature–enthalpy coordinates, and Figure 6.9.1c shows temperature changes at inlets and outlets of individual exchangers.

In the network the high-temperature heat source, having power of 200 kW, is realized by supplying live steam. The network needs to be modernized so that the consumption of live steam is reduced without changing the temperatures or mass flows at inlets and outlets of the exchanger network. The proposed modernization is to utilizing a vapour compression cycle heat pump.

The modernized heat exchanger network is shown in Figures 6.9.2a, 6.9.2b and 6.9.2c, corresponding to the original figures. As a result of using the heat pump, there is a need to introduce an additional heat exchanger, denoted by number 5.

Calculations of the approximate cycle for the proposed heat pump were performed assuming thermal power in the low-temperature and high-temperature heat sources as 100 and 150 kW, respectively. Because of the high-temperature of the high-temperature source, R123 was adopted as the working fluid. The intensive parameters of the cycle are presented in Table 6.9.1. Electrical power consumed by

Table 6.9.1

	T (K)	P (MPa)	i (kJ/kg)	s (kJ/kg K)
1	330.15	0.264	238.0	3.86
2	406.15	1.212	285.0	3.91
2'	393.15	1.212	273.8	3.88
3	393.15	1.212	151.8	2.69
3'	393.15	1.212	151.6	2.69
4	330.15	0.264	151.8	2.78

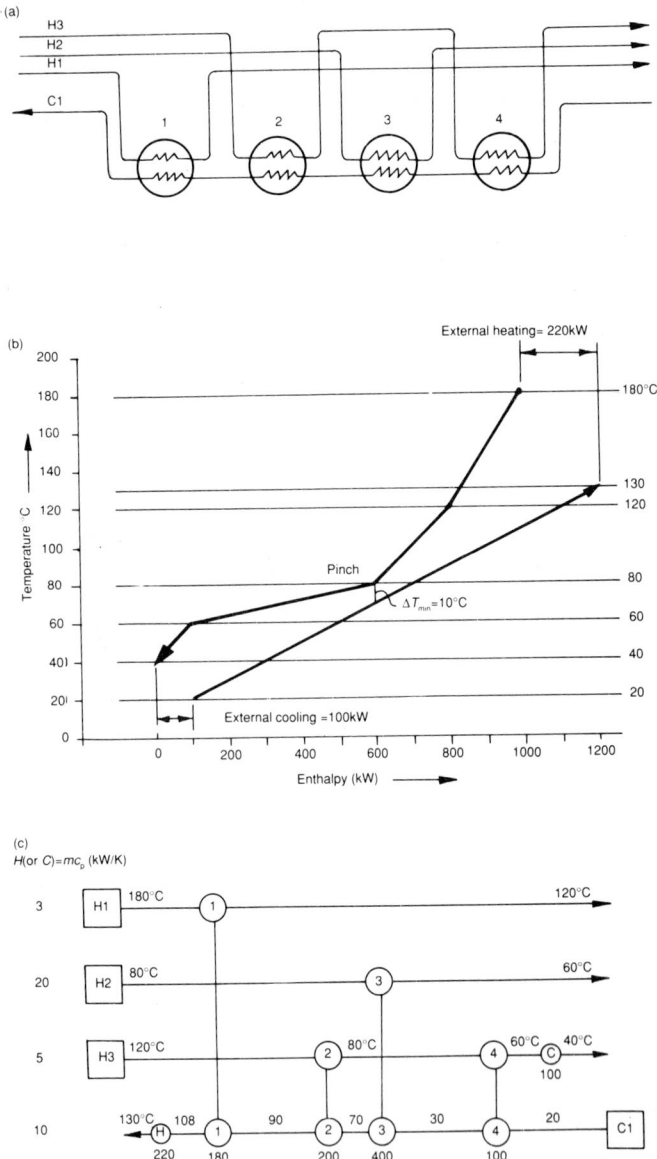

Figure 6.9.1   *Characteristics of a thermal network consisting of four heat exchangers, proposed for modernization: (a) schematic diagram of the network with media flowing through the exchangers; (b) temperature changes of the cooled and heated media. Values of heat fluxes supplied in the external high-temperature source and discharged in the external low-temperature source are also given*

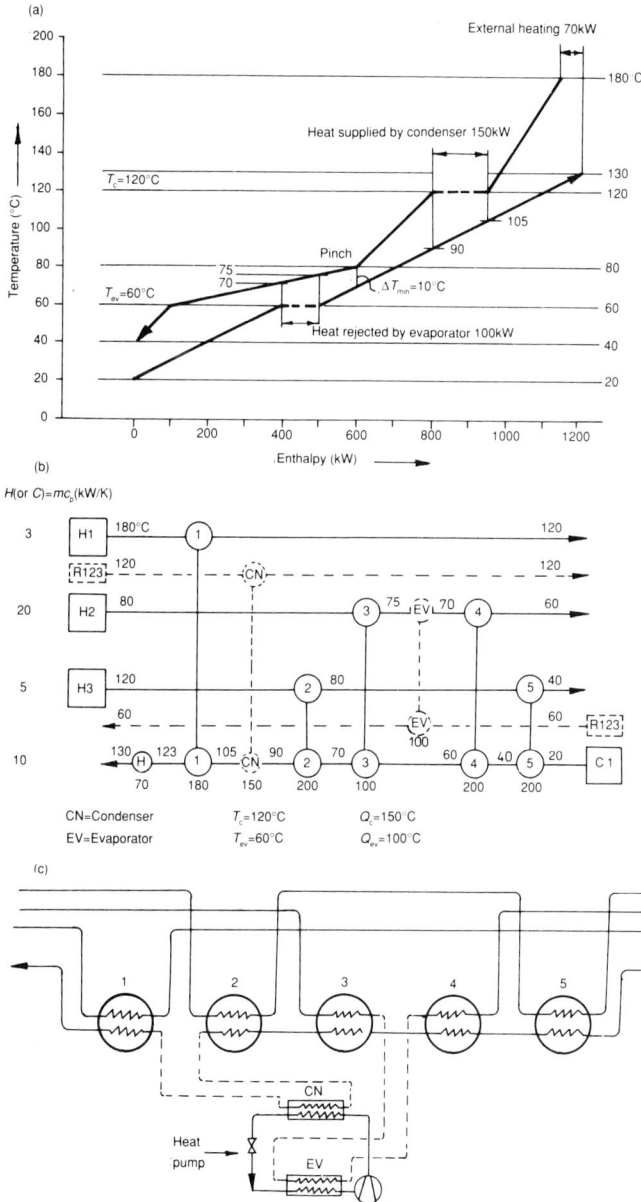

Figure 6.9.2 *Characteristics of the modernized heat exchanger network;* (a) *schematic diagram of the installation supplemented by a heat pump and an additional exchanger (number 5);* (b) *changes in temperatures of the flowing media, taking into account the presence of the heat pump working fluid and reduced requirement for thermal energy at the high-temperature source;* (c) *temperatures of the media at inlets and outlets from the heat exchangers*

the motor driving the piston compressor, having isentropic efficiency of 0.7, is 62.5 kW (taking into account electrical losses at the efficiency of 0.8).

As a result of the proposed modernization, the requirement for energy from an external source has been reduced by 150 kW, and additionally the need to discharge heat into an external low-temperature source has been totally eliminated. Because of the high-temperature (180°C) of the external high-temperature source the proposed modernization does not provide total elimination of the live steam demand.

Appendices

A.1 Formulae for computing thermodynamic parameters of R12 and R22 working fluids

The enthalpy of liquid and vapour at the saturation line, according to [125], is defined by the following formulae.

For the liquid:

$$i_L = a_0 + a_1(T_s - 273.15) \text{ kJ/kg} \tag{A.1}$$

For the vapour:

$$i_V = b_0 + b_1(T_S - 273.15) \text{ kJ/kg} \tag{A.2}$$

The increase of enthalpy during isentropic compression, according to [125] is defined by the following formula:

$$\Delta i_{is} = \alpha(T_1 - T_2) \text{ kJ/kg} \tag{A.3}$$

where $\alpha = c_0 + c_1(T_2 - 273.15)$

The pressure at the saturation line is defined by the formula:

$$P_S = d_0 + d_1(T_S - 273.15) + d_2[(T_S - 273.15)^2 + d_3(T_S - 273.15)^3] \text{ MPa} \tag{A.4}$$

Values of the coefficients appearing in these formulae, for R12 and R22, are given in Table A.1.

Table A.1 *Values of coefficients in Equations A.1 to A.4*

Coefficient	Units	Working fluid R12	Working fluid R22
a_0	kJ/kg	199.790	200.600
a_1	kJ/(kg K)	1.020	1.340
b_0	kJ/kg	355.190	407.020
b_1	kJ/(kg K)	0.378	0.280
c_0	kJ/(kg K)	0.473	0.680
c_1	kJ/(kg K^2)	-0.166×10^{-2}	-0.47×10^{-2}
d_0	MPa	3.148×10^{-1}	5.105×10^{-1}
d_1	MPa/K	1.01×10^{-2}	1.67×10^{-2}
d_2	MPa/K^2	1.256×10^{-4}	2.049×10^{-4}
d_3	MPa/K^3	0.747×10^{-6}	1.51×10^{-6}
Temperature range	K	263–333	263–323

A.2 Formulae for computing thermodynamic parameters of working fluid and sorbent pairs

A.2.1 R717–water

Thermodynamic parameters of the pair ammonia–water can be estimated using functions defining free enthalpies of the working fluid and sorbent and their mixtures in the liquid and vapour phases. These functions, for the pressure range 0.02–5.0 MPa and temperature range 230–500 K, are defined by Equations A.5 to A.8, quoted after Ziegler [126]. Dimensionless parameters appearing in these formulae are defined as follows:

$$g_R = \frac{g}{RT_B}, \quad h_R = \frac{h}{RT_B}, \quad P_R = P/P_B, \quad T_R = T/T_B$$

and the values of constants P_B, R, T_B are, respectively:

$$P_B = 1\,\text{MPa}, \quad R = 8.314\,\text{kJ/(kmol K)}, \quad T_B = 100\,\text{K}$$

Free enthalpies of ammonia and water are defined by the following formulae.
For the liquid phase:

$$g_R^l = h_{oR}^l - T_R s_{oR}^l + \int_{T_{oR}}^{T_R} c_{PR}^l \, dT_R + T_R \int_{T_{oR}}^{T_R} (c_{PR}^l/T_R) \, dT_R +$$

$$+ (A_1 + A_3 T_R + A_1 T_R^2)(P_R - P_{oR}) + A_2(P_R^2 - P_{oR}^2)/2 \tag{A.5}$$

For the vapour phase:

$$g_R^g = h_{oR}^g - T_R s_{oR}^g + \int_{T_{oR}}^{T_R} c_{PR}^g \, dT_R + T_R \int_{T_{oR}}^{T_R} (c_{PR}^g/T_R) \, dT_R + \ln(P_R/P_{oR}) +$$

$$+ C_1(P_R - P_{oR}) + C_2(P_R/T_R^3 - 4P_{oR}/T_{oR}^3 + 3P_{oR}T_R/T_{oR}^4) +$$

$$+ C_3(P_R/T_R^{11} - 12P_{oR}/T_{oR}^{11} + 11P_{oR}T_R/T_{oR}^{12}) +$$

$$+ C_4(P_R^3/T_R^{11} - 12P_{oR}^3/T_{oR}^{11} + 11P_{oR}^3 T_R/T_{oR}^{12})/3 \tag{A.6}$$

where:

$$c_{PR}^l = B_1 + B_2 T_R + B_3 T_R^2,$$
$$c_{PR}^g = D_1 + D_2 T_R + D_3 T_R^2$$

Values of these coefficients, for ammonia and water, are given in Table A.2.
The free enthalpy of the mixture is defined by the following formulae:

For the liquid phase:

$$g_{RM}^l = (1 - x)g_{R(H_2O)}^l + xg_{R(NH_3)}^l + T_R[(1 - x)\ln(1 - x) + x \ln x]$$

$$+ \{E_1 + E_2 P_R + (E_3 + E_4 P_R)T_R + E_5/T_R + E_6/T_R^2 +$$

$$+ [E_7 + E_8 P_R + (E_9 + E_{10} P_R)T_R + E_{11}/T_R + E_{12}/T_R^2](2x - 1) +$$

$$+ [E_{13} + E_{14} P_R + E_{15}/T_R + E_{16}/T_R^2](2x - 1)^2\} x(1 - x) \tag{A.7}$$

Table A.2 *Values of coefficients in Equations A.5 and A.6*

Coefficients	Ammonia	Water
A_1	3.971423×10^{-2}	2.748796×10^{-2}
A_2	-1.790557×10^{-5}	-1.016665×10^{-5}
A_3	-1.308905×10^{-2}	-4.452025×10^{-3}
A_4	3.752836×10^{-3}	8.389246×10^{-4}
B_1	1.634519×10^{1}	1.214557×10^{1}
B_2	-6.508119	-1.898065
B_3	1.448937	2.911966×10^{-1}
C_1	-1.049377×10^{-2}	2.136131×10^{-2}
C_2	-8.288224	-3.169291×10^{1}
C_3	-6.647257×10^{2}	-4.634611×10^{4}
C_4	-3.045352×10^{3}	0.0
D_1	3.673647	4.019170
D_2	9.989629×10^{-2}	-5.175550×10^{-2}
D_3	3.617622×10^{-2}	1.951939×10^{-2}
h_{oR}^{l}	4.878573	21.821141
h_{oR}^{g}	26.468879	60.965058
s_{oR}^{l}	1.644773	5.733498
s_{oR}^{g}	8.339026	13.453430
T_{oR}	3.225200	5.070500
P_{oR}	2.000000	3.000000

Table A.3 *Values of coefficients in Equations A.7 and A.8*

E_1	-4.626129×10^{1}
E_2	2.060225×10^{-2}
E_3	7.292369
E_4	-1.032613×10^{-2}
E_5	8.074824×10^{1}
E_6	-8.461214×10^{1}
E_7	2.452882×10^{1}
E_8	9.598767×10^{-3}
E_9	-1.475383
E_{10}	-5.038107×10^{-3}
E_{11}	-9.640398×10^{1}
E_{12}	1.226973×10^{2}
E_{13}	-7.582637
E_{14}	6.012445×10^{-4}
E_{15}	5.487018×10^{1}
E_{16}	-7.667596×10^{1}

For the vapour phase:

$$g_{RM}^g = (1 - y)g_{R(H_2O)}^g + yg_{R(NH_3)}^g + T_R[(1 - y)\ln(1 - y) + y\ln y] \tag{A.8}$$

where x = molar fraction of ammonia in the mixture in the liquid phase, y = molar fraction of ammonia in the mixture in the gas phase.

The values of coefficients $E_1, ..., E_{16}$ are given in Table A.3.

The formulae are derived as a generalization of results presented by Schulz [65, 127].

A.2.2 R18–water solution of lithium bromide

The boiling point of the water solution of lithium bromide $T_{bo,so}$ (following the Duhring rule), according to [128], is given by the following formula:

$$T_{bo,so} = (a_0 + a_1x + a_2x^2 + a_3x^2)T_{bo,w} + (b_0 + b_1x + b_2x^2 + b_3x^3) \tag{A.9}$$

where $T_{bo,w}$ = boiling point for water at the same pressure as the solution, and the coefficients are given in Table A.4.

Table A.4 *Values of coefficients in Equation A.9*

a_0	-2.007550
a_1	0.169760
a_2	-3.13336×10^{-3}
a_3	1.97668×10^{-5}
b_0	124.937000
b_1	-7.716500
b_2	-0.152286
b_3	-7.9509×10^{-4}

This formula is valid in the following ranges:

$278.15\,\text{K} < T_{bo,so} < 458.15\,\text{K}$; $273.15\,\text{K} < T_{bo,w} < 383.15\,\text{K}$; $0.45 < x < 0.70$

The specific enthalpy of the water solution of lithium bromide at the saturation line is defined by the following formula, after [128]:

$$i = 2.326(A + B)[1.8(T_{bo,so} - 273.15) + 32] + C[T_{bo,so} - 273.15)^2]\ \text{kJ/kg} \tag{A.10}$$

The coefficients appearing in this formula can be calculated from the following:

$$A = -1015.07 + 79.5387x - 358\,016x^2 + 0.03031583x^3 - (1.400261 \times 10^{-4})x^4$$

$$B = 4.68108 - (3.037766 \times 10^{-1})x + (8.44845 \times 10^{-3})x^2 -$$
$$- (1.047721 \times 10^{-4})x^3 + (4.80097 \times 10^{-7})x^4$$

$$C = (-4.9107 \times 10^{-3}) + (3.83184 \times 10^{-4})x - (1.078963 \times 10^{-5})x^2 +$$
$$+ (1.3152 \times 10^{-7})x^3 - (5.897 \times 10^{-10})x^4 \tag{A.11}$$

The functions describing other properties of the pair R718–water solution of lithium bromide are given, among others, in [129].

A.3 log P–i charts

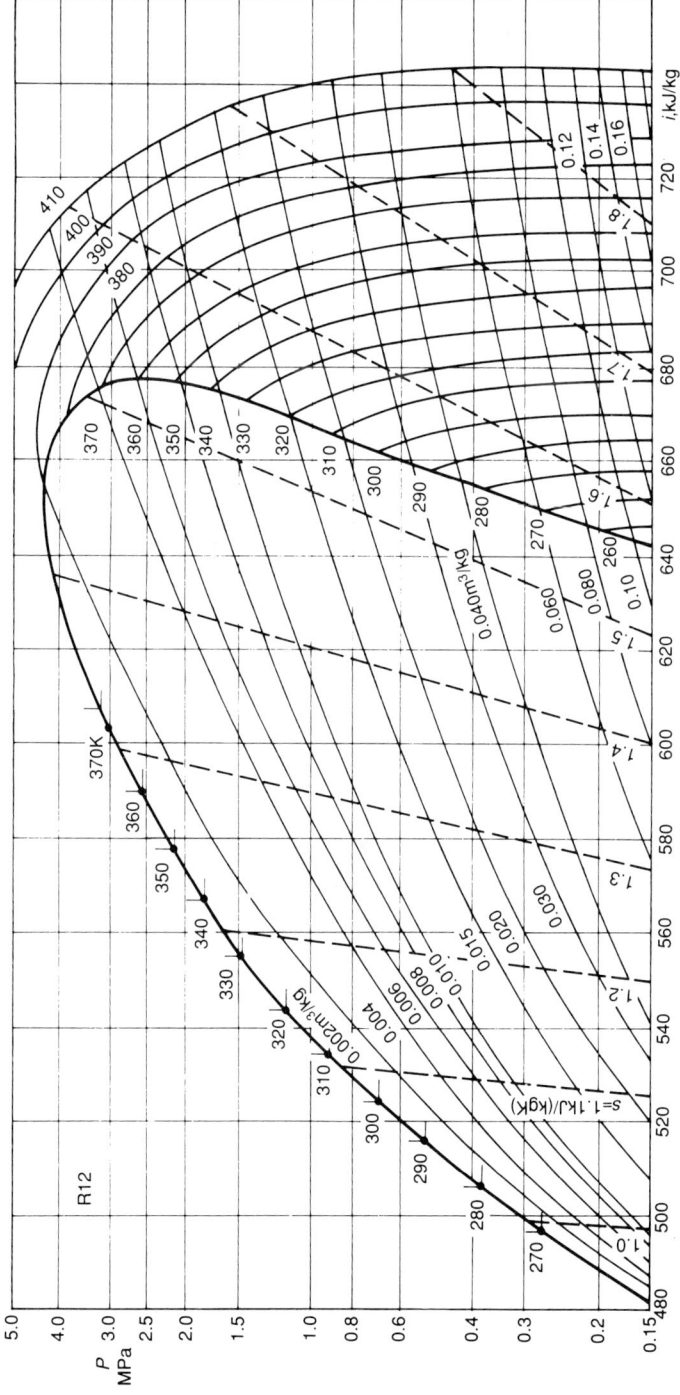

Figure A.1 *log P–i chart for the R12 fluid; standard conditions for liquid: T = 273 K, i = 500 kJ/kg, s = 1 kJ/(kg K)*

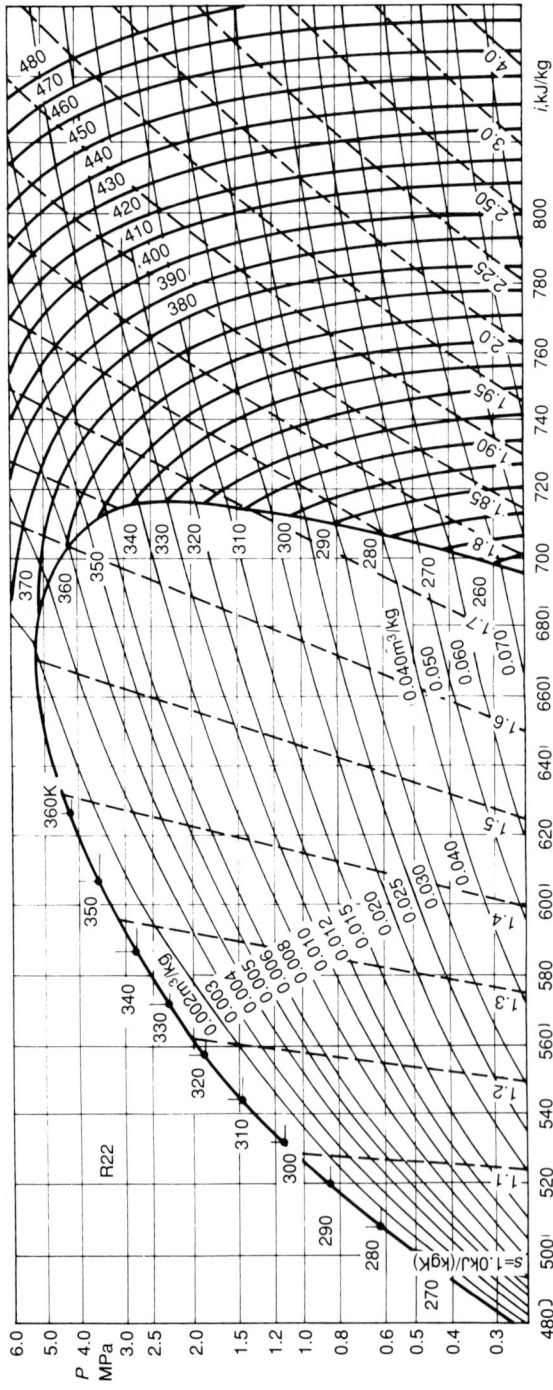

Figure A.2 *log P–i chart for the R22 fluid; standard conditions for liquid: T = 273 K, i = 500 kJ/kg, s = 1 kJ/(kg K)*

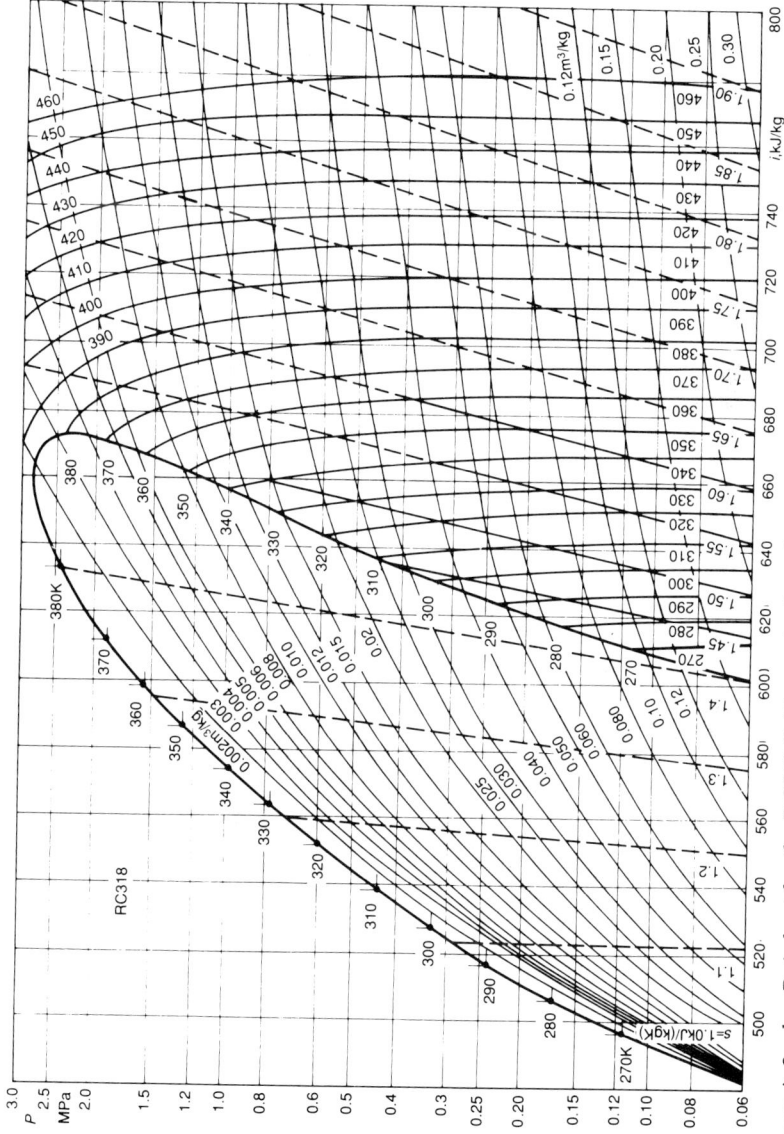

Figure A.3 *log P–i chart for the RC318 fluid; standard conditions for liquid:* $T = 273\ K$, $i = 500\ kJ/kg$, $s = 1\ kJ/(kg\ K)$

A.4 *i–z* charts

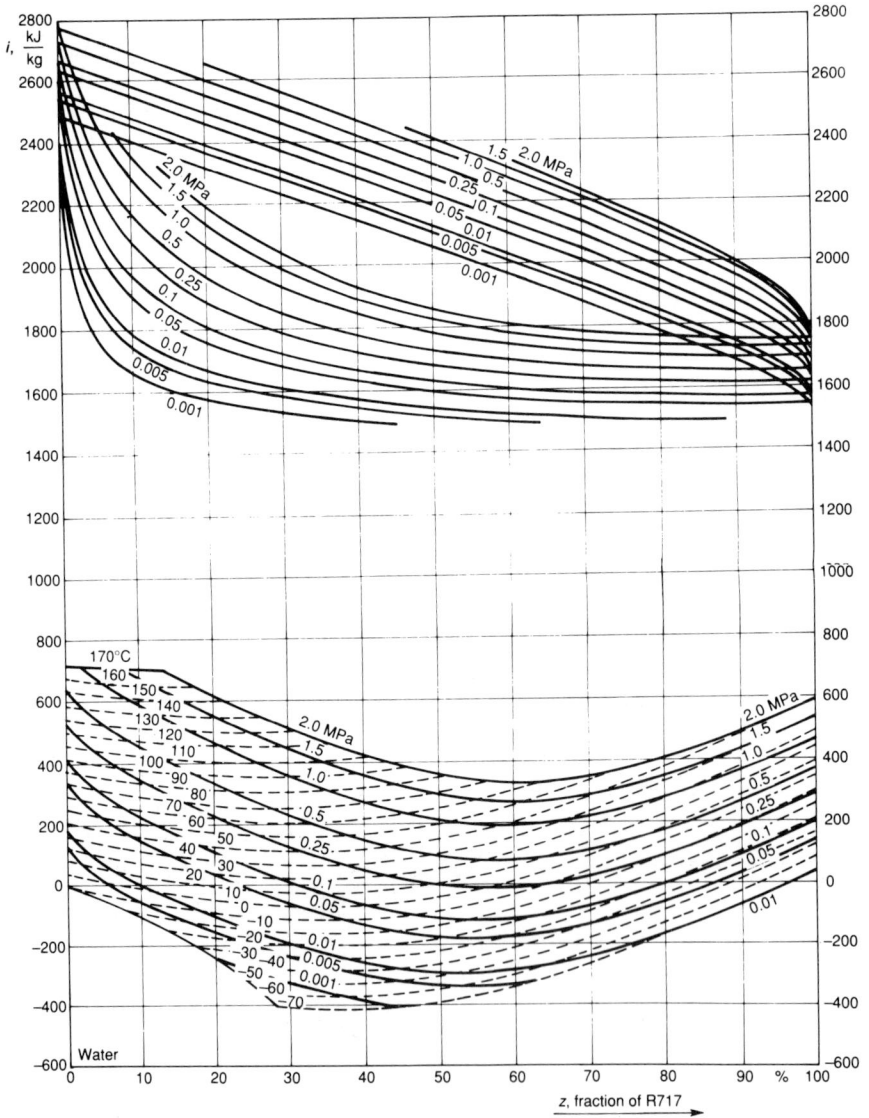

Figure A.4   *i–z chart for the R717 – water pair of fluids; standard conditions: enthalpy of components at the liquid line for T = 273 K equal zero*

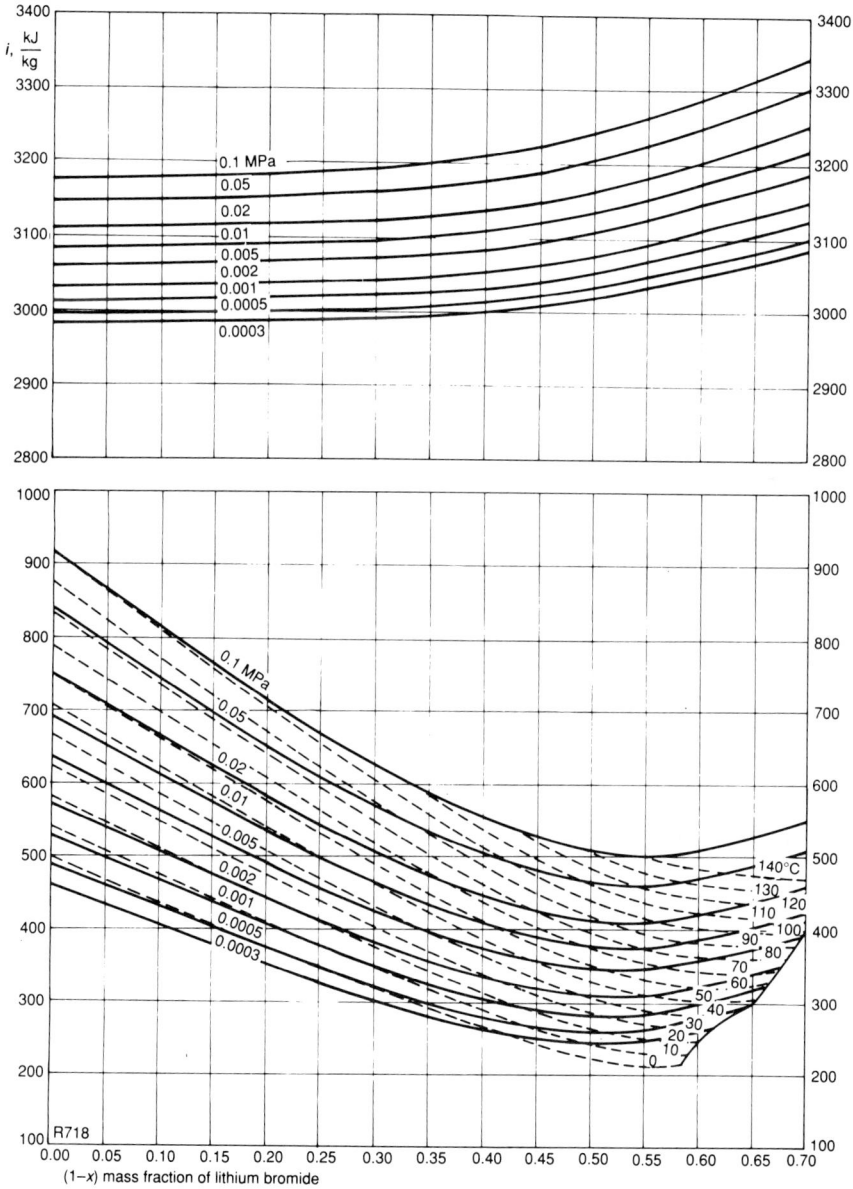

Figure A.5 *i–z chart for the R718–water solution of lithium bromide pair of fluids (simplified chart neglecting the presence of lithium bromide vapours in the working fluid); standard conditions: enthalpy of water at the liquid line for T = 273 K equal zero*

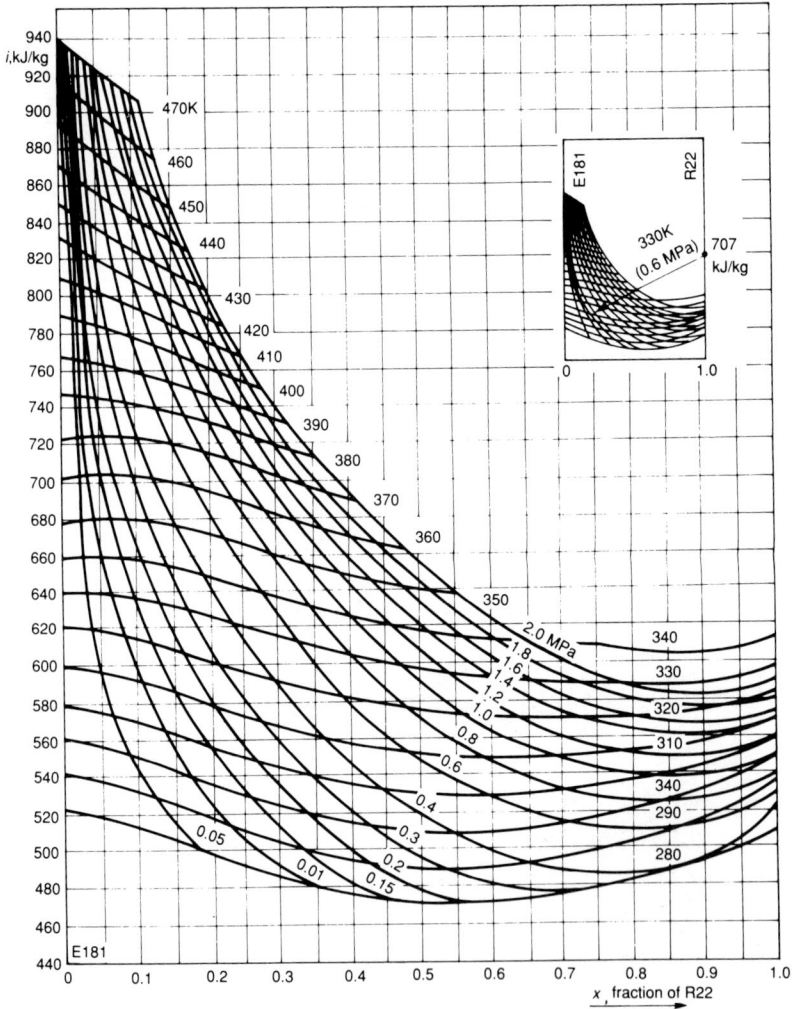

Figure A.6   *i–z chart for the R22–E181 pair of fluids (simplified chart neglecting the presence of sorbent vapours in the working fluid); standard conditions: enthalpy of components at the liquid line for T = 273 K: for z = 0 and z = 1; i = 500 kJ/kg*

A.5 log *P–z* charts

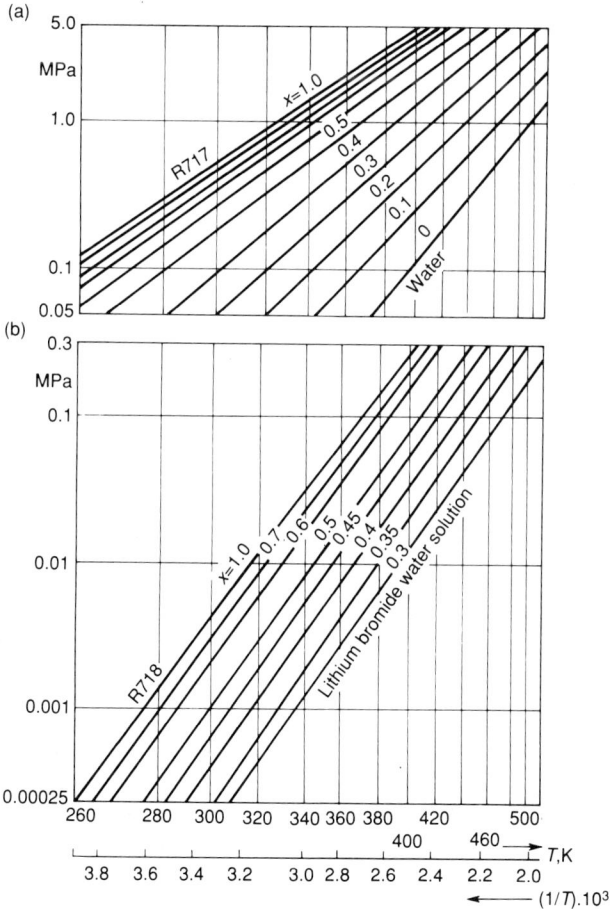

Figure A.7 *log P–i charts for:* (a) *the R717–water pair of fluids;* (b) *R718–lithium bromide*

A.6 Performance data of mechanical vapour compression cycle heat pumps

Table A.5 *Mechanical vapour compression cycle heat pumps; extract from the catalogue of STAL Refrigeration AB, Norrköping, Sweden*

Model	Mass (kg)	Noise level (dB)	Installed drive power (kW)	T_{ls}/T_{hs} (°C/°C)	Working fluid	Thermal power Q_{hs} (kW)	COP
VDP 2	550	71	11	+10/+50	R22	52	3.60
VDP 2	550	71	11	+10/+55	R12	32	3.60
VDP 3	600	72	19	−10/+40	R22	41	3.20
VDP 3	600	72	19	−10/+55	R12	22	2.30
VDP 4	670	74	26	+10/+55	R22	109	3.20
VDP 4	670	74	26	+10/+70	R12	57	2.90
VDP 4	670	74	26	+20/+70	R12	80	3.40
VMP 106	1500	79	45	−10/+40	R22	106	3.23
VMP 106	1500	79	45	+10/+80	R114	31	2.06
VMP 112	2400	82	90	+10/+70	R12	206	2.83
VMP 124	3450	85	180	−10/+70	R12	187	2.11
VMP 124	3450	85	180	−10/+55	R500	249	2.53
VMP 124	3450	85	180	+5/+40	R22	685	4.03
VSP 51	6235	91	160	−10/+40	R22	545	3.30
VSP 51	6235	91	160	−10/+70	R12	310	1.88
VSP 73 E	14 000	98	585	−10/+40	R22	1980	3.46
VSP 73 E	14 000	98	585	−10/+70	R12	1330	2.18
VSP 73 E	14 000	98	585	+10/+100	R114	800	2.16
VSP 73 E	14 000	98	585	+10/+40	R22	3420	4.87
VKP 93	30 000	96	1150	−10/+40	R22	3690	3.25
VKP 93	30 000	96	1150	−10/+55	R12	2180	2.49
VKP 93	30 000	96	1150	+10/+70	R114	1350	2.56
VKP 93	30 000	96	1150	+10/+90	R114	1160	1.75
VKP 93 E	33 000	96	1350	−10/+40	R22	4490	3.46
VKP 93 E	33 000	96	1350	+10/+40	R22	7800	4.42
VKP 93 E	33 000	96	1350	−10/+70	R12	2840	2.20
VKP 93 E	33 000	96	1350	+30/+100	R114	3100	2.47
VKP 93 E	33 000	96	1350	+40/+100	R114	3853	2.85

A.7 Examples of design solutions of sorption cycle heat pumps

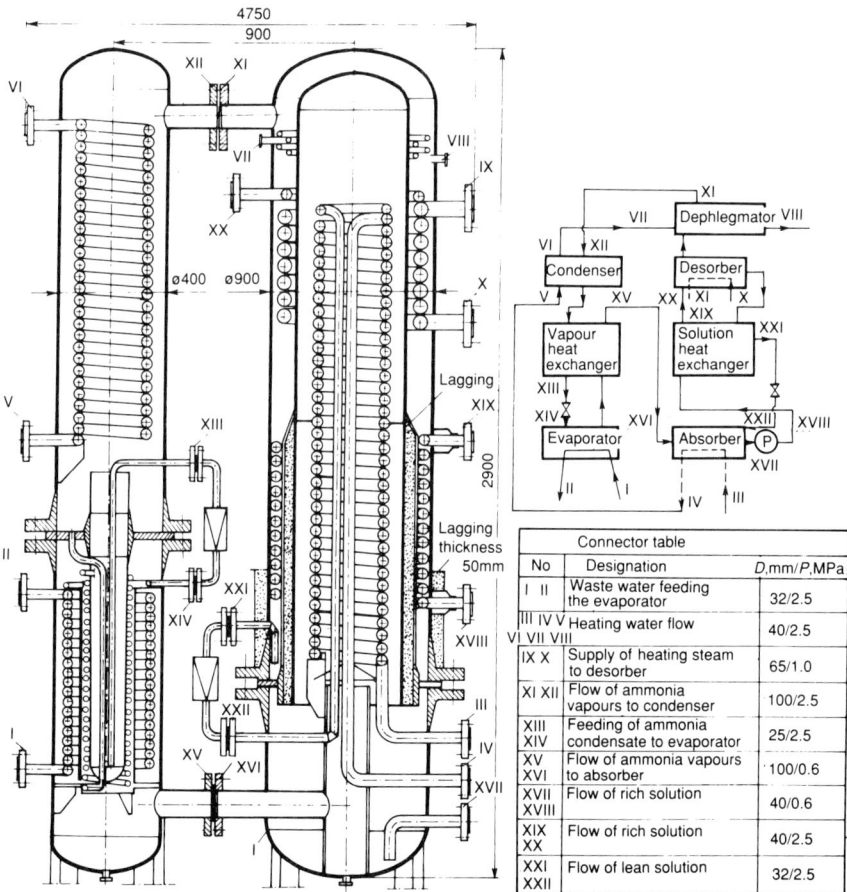

4750
900

XII XI
VI
VIII
VII
IX
XX
ø400 ø900
X

XI
VII Dephlegmator VIII
VI XII
Condenser   Desorber
V XV XX XI X
XIX
Vapour   Solution
heat   heat   XXI
exchanger   exchanger
Lagging
XIX
XIII
XIV XVI
V
XIII
Evaporator   Absorber (P) XVIII
XVII
II
2900
Lagging
thickness
50mm

II
XXI
XIV
XVIII
XXII
III
I
XV XVI
IV
XVII
I

Connector table		
No	Designation	D,mm/P,MPa
I II	Waste water feeding the evaporator	32/2.5
III IV V VI VII VIII	Heating water flow	40/2.5
IX X	Supply of heating steam to desorber	65/1.0
XI XII	Flow of ammonia vapours to condenser	100/2.5
XIII XIV	Feeding of ammonia condensate to evaporator	25/2.5
XV XVI	Flow of ammonia vapours to absorber	100/0.6
XVII XVIII	Flow of rich solution	40/0.6
XIX XX	Flow of rich solution	40/2.5
XXI XXII	Flow of lean solution	32/2.5

Figure A.8 *Diagrammatic representation of a design of a high-power absorption cycle heat pump*

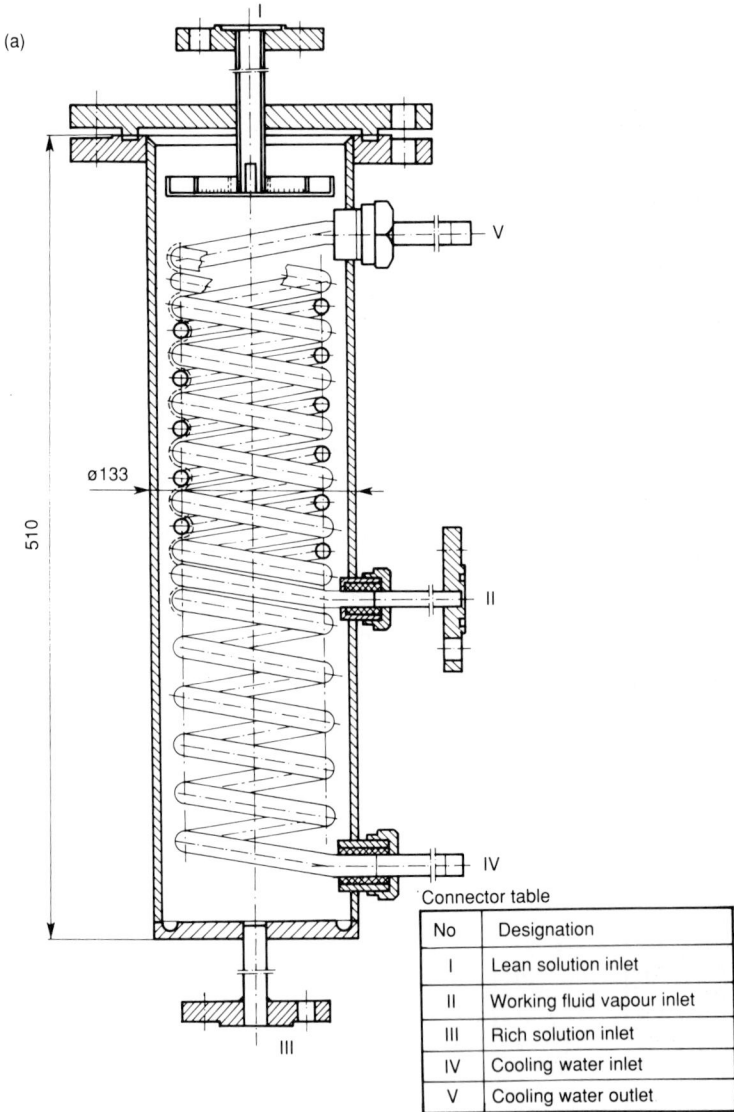

(a)

ø133

510

No	Designation
I	Lean solution inlet
II	Working fluid vapour inlet
III	Rich solution inlet
IV	Cooling water inlet
V	Cooling water outlet

Connector table

Figure A.9   *Exchangers of a low-power heat pump:* (a) *absorber;* (b) *desorber*

(b)

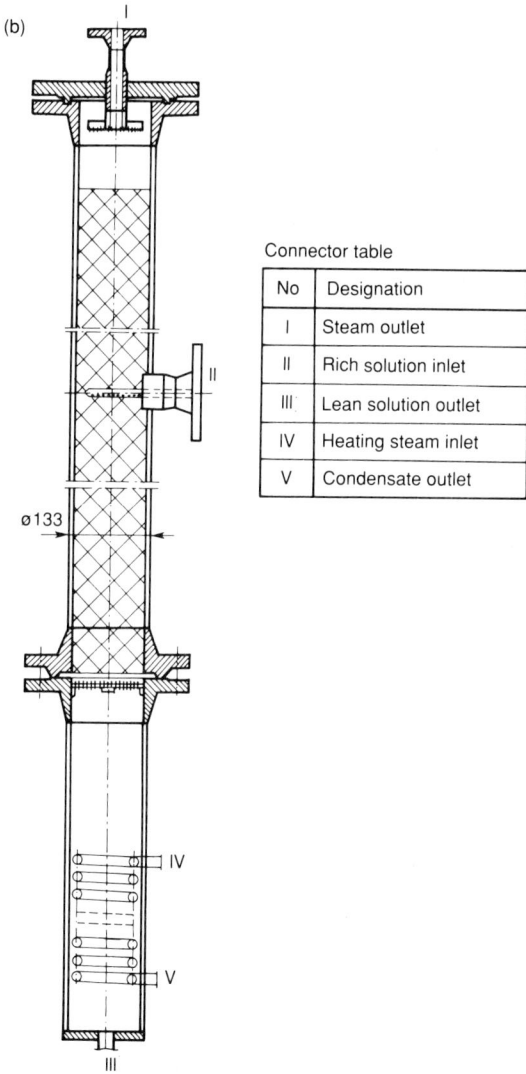

Connector table

No	Designation
I	Steam outlet
II	Rich solution inlet
III	Lean solution outlet
IV	Heating steam inlet
V	Condensate outlet

ø133

Figure A.9   *(Continued)*

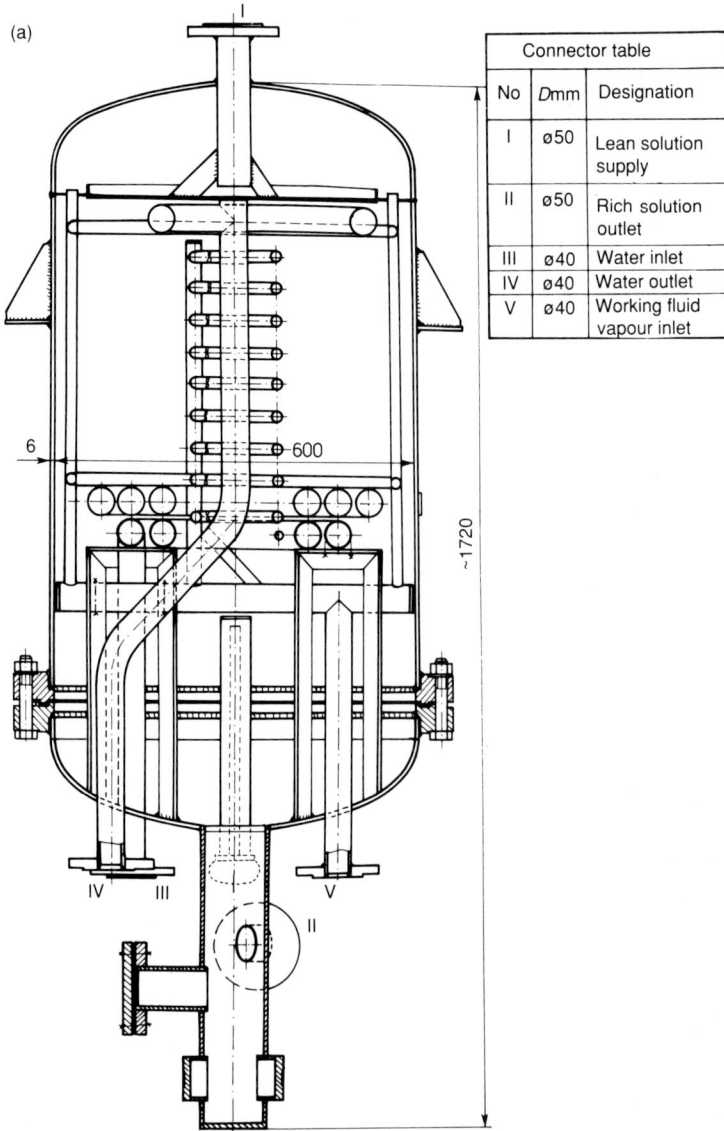

(a)

Connector table		
No	Dmm	Designation
I	ø50	Lean solution supply
II	ø50	Rich solution outlet
III	ø40	Water inlet
IV	ø40	Water outlet
V	ø40	Working fluid vapour inlet

Figure A.10   *Medium-power heat exchangers:* (a) *absorber;* (b) *desorber*

(b)

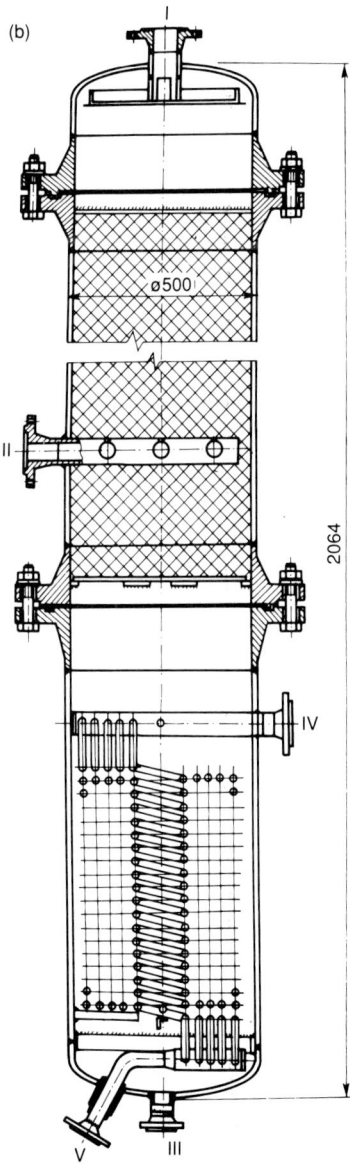

Connector table		
No	*D*mm	Designation
I	ø65	Ammonia vapour outlet
II	ø50	Rich solution inlet
III	ø50	Lean solution outlet
IV	ø50	Steam inlet
V	ø30	Condensate outlet

ø500

2064

Figure A.10   *(Continued)*

References

1. Haldane, J. G. N. (1930) The heat pump – an economical method of producing low grade heat from electricity. *IEE Journal*, pp. 666–675.
2. Kernan, G. and Brady, J. (1977) Economic evaluation of heat pumps. *Int. J. Energy Res.,* No. 2, pp. 115–125.
3. Cacciola, G. *et al.* (1987) Chemical heat pump using heat of reversible catalytic reactors. *Int. J. Energy Res.*, No. 9, pp. 519–529.
4. Clarke, E. C. and Morgan, O. (1983) Chemical heat pumps for industry. Proc. 16th Intersoc. Energy Convers. Eng. Conf. USA, 1983.
5. Fujiwara, I.C. (1982) Chemical heat pump system. *Zairyo: Kogyo Zairyo,* pp. 69–72.
6. Greiner, L. (1984) Chemical heat pump. US Patent 4, 425, 903.
7. Raldow, M. W. and Wentworth, W. E. (1979) Chemical heat pumps. A basic thermodynamic analysis. *Solar Energy*, No. 1, pp. 75–80.
8. Saito, Y., Kameyame, H. and Yoshida, K. (1987) Catalyst-assisted chemical heat pump with reaction couple of acetone hydrogeneration/2-propared dehydrogenation for upgrading low level thermal energy: proposal and evaluation. *Int. J. Energy Res.,* No. 3, pp. 549–558.
9. Wentworth, W. E., Johnston, D. A. and Raldow, W. M. (1978) Chemical heat pump using a slurry of metal salt ammoniates in an inert solvent. Workshop on Chemical Heat Pump Technology, Iceland, 1978.
10. Bonc, K., Stencel, B. and Wysokinski, S. (1980) Efekt Ranque'a-Hilscha. *Chlondnictwo*, No. 5, pp. 5–9.
11. Jeter, M. S. (1987) A concept for an innovative high-temperature heat pump. *Energy*, No. 4, pp. 163–169.
12. Alefeld, G. (1987) A high temperature absorption heat pump as topping process for power generation. *Energy*, No. 3, pp. 649–659.
13. Alefeld, G. (1981) Metal hydrides as heat transformer energy cascading topics. *Appl. Phys.*, No. 29, pp. 501–520.
14. Argabright, T. A. (1982) Metal hybride-chemical heat pump development project, Phase I. Final Report BNL Rep. 51539.
15. Ferens, B. and Rubik, M. (1972) *Chlodzenie Termoelektryczne w Klimatyzacji.* Warszawa: Arkady, pp. 189–266.
16. Szargut, J. and Petela, R. (1965) *Egzergia.* Warszawa: WNY.
17. Staniszewski, B. (1986) *Termodynamika.* Warszawa: PWN.
18. Wisniewski, S. (1980) *Termodynamika Techniczna.* Warszawa: WNT.
19. Von Cube, H. L. (1980) Warmequellen für Warmepumpen. In: *Warmepumpentechnologie*, Vol. II, Essen: Vulkan-Verlag, pp. 173–181.
20. Borel, L. (1976) Energy economics and energy comparison of different heating systems based on the theory of exergy. In: *Heat Pumps and their Contribution to Energy Conservation* (ed. Camatini, E. and Kester T.), Noordhoff: Leyden, pp. 51–95.
21. DuPont Company (1991) Alternatives to chlorofluorocarbons (personal communication), DuPont Freon Products Laboratory, Wilmington, DE 19898.
22. Statt, T. G. (1990) An overview of ozone-safe refrigerants for centrifugal chillers. *ASHRAE Trans.*, Vol. 96, Part 2, pp. 1424–1428.

23. Kruse, H. and Hesse, U. (1988) Possible substitutes for fully halogenated chlorofluorocarbons using fluids already marketed. *Int. J. Refrig.*, Vol. 11, pp. 276–283.
24. Kruse, H. (1984) Improving industrial heat pumps by applying refrigerant mixtures. *Heat Recovery Systems*, No. 5, pp. 359–363.
25. Degueurce, B. (1984) Use of twin screw compressors for steam compression. Proc. Int. Symposium on the Large Scale Applications of Heat Pumps, England, pp. 189–196.
26. Iyoki, S. and Vamura, T. (1978) Studies on corrosion inhibitors in water–lithium bromide absorption refrigerating machines. *Resto*, pp. 1101–1109.
27. Eisa, M. A. R. and Holland, F. A. (1987) A study of the optimum interaction between the working fluid and the absorbent in absorption heat pump systems. *Heat Recovery Systems*, No. 4, pp. 107–117.
28. Eisa, M. A. R., Best, R. and Holland, F. A. (1986) Working fluids for high temperature heat pumps. *Heat Recovery Systems*, No. 1, p. 305.
29. Gazinski, B. (1983) Roztwory robocze absorpcyjnych pomp ciepla. *Chlodnictwo*, No. 2, pp. 8–13.
30. Iedema, P. D. (1982) Mixtures for the absorption heat pump. *Int. J. Refrig.*, No. 8, pp. 262–268.
31. Stephan, K. and Seher, D. (1980) Arbeitsgemische fur Sorptions-Warmepumpen. *Kalte-Klima-Heizung*, No. 2, pp. 865–876.
32. Von Cube, H. L. and Steimle, F. (1978) *Warmepumpen, Grundlagen und Praxis*, Essen: Vulkan-Verlag.
33. *Miesieczny Biuletyn Hydrologiczno-Meteorologiczny*, from No. 2, 1983 to No. 1, 1984.
34. *Rocznik Hydrologiczny wod Powierzchniowych*, 1977 (1982) Warszawa: WKiL.
35. *Rocznik Hydrologiczny wod Powierzchniowych*, 1978 (1982) Warszawa: WKiL.
36. Plochniewski, Z. (1978) Warunki wystepowania wod termalnych na obszarze Polski oraz mozliwosci ich uzyskania i zastosowania. *Problemy Uzdrowiskowe*, Vol. 3(125), pp. 97–103.
37. Neill, D. T. and Jensen, W. P. (1976) Geothermal powered heat pumps to produce process heat. Proc. 11th Intersoc. Energy Conv. Eng. Conf. USA, 1976.
38. Gogol, W., Gogol, E. and Artecka, L. (1973) Badania przewodnosci cieplnej gruntow wilgotnych. *Biul. Inform. ITC PW*, No. 54, pp. 49–69.
39. Svec, V. I. (1987) Potential of ground heat source systems. *Int. J. Energy Res.*, No. 2, pp. 573–581.
40. Jaskolski, K. (1981) *Struktura Promieniowania Slonecznego w Polsce.* Sprawozdanie Instytutu Energetyki.
41. Akalin, M. T. (1978) Equipment life and maintenance cost survey. *ASHRAE Trans.*, (1978) No. 4, pp. 315–320.
42. Patwardhan, V. R. and Patwardhan, V. S. (1987) A simplified procedure for the estimation of COP for heat pumps. *Heat Recovery Systems*, No. 5, pp. 435–450.
43. Ziegler, F. and Alefeld, G. (1987) Coefficient of performance of multistage absorption cycles. *Rev. Int. Froid.*, No. 1, pp. 285–295.
44. Trommelmans, J., van-den Bulck, E. and Berghmans, J. (1981) Factors influencing the performance of a domestic absorption heat pumps. Proc. Int. Conference of Refrigeration, Essen RFN, 1981.
45. Angelino, G. (1976) Development of thermal prime movers for heat pump drive. In: *Heat Pumps and their Contribution to Energy Conservation* (ed. Camatini, E. and Kester, T.), Noordhoff: Leyden, pp. 155–200.
46. Fordsmand, M. (1978) Analysis of the factors which determine the COP of a heat pump, and a feasibility study on ways and means of increasing same. Proc: EEC Contractors Meeting on Heat Pumps, Brussels, 1978.
47. Hummel, W. S. (1980) Der Dampfstrahl-Verdichter als Warmepumpe. In: *Warmepumpentechnologie*, Vol. I. Essen: Vulkan-Verlag.

48. Macchi, E. (1983) Prime movers for vapour compression heat pumps. In: *Heat Pump Fundamentals* (ed. Berghmans J.), London: Martinus-Nijhoff Publishers, pp. 192–227.

49. Macmichael, D. B. A., Reay, D. A. and Scarle, N. K. (1978) Feasibility and design study of a gas engine driven high temperature industrial heat pump. Proc. EEC Contractors Meeting on Heat Pumps, 1978.

50. Shearer, A. (1976) Selection of prime movers for on site energy generation. *Power Generation Industrial*, pp. 26–30.

51. Eisa, M. A. R. (1986) Classified reference for absorption heat pump systems from 1975 to May 1985. *Heat Recovery Systems*, No. 1, pp. 47–61.

52. Gajczak, S. (1956) *Absorpcyjne Urzadzenia Chlodnicze*. Warszawa: WNT.

53. Glaser, H. (1980) Thermodynamische Grundlagen der Absorptionswarmepumpen. In: *Warme-pumpentechnologie*, Vol. II, Essen: Vulkan-Verlag, pp. 62–71.

54. Glaser, H. (1982) Ein praktische Vergleichsprozess für die Absorptionwarmepumpen *Ki Klima-Kaltze-Heizung*, No. 6, pp. 233–240.

55. Best, R., Eisa, M. A. R. and Holland, F. A. (1987) Thermodynamics design data for absorption heat pump system operating on ammoniawater – Part II Heating. *Heat Recovery Systems*, No. 2, pp. 177–185.

56. Grossman, G. (1982) Adiabatic absorption and desorption for improvement of temperature-boosting absorption heat pump. *ASHRAE Trans.*, No. 6, pp. 359–367.

57. Grossman, E. R. and Shavrin, V. S. (1982) Increasing effectiveness of absorption heat pumps. *Kholod Tech.*, No. 1, pp. 28–35.

58. Akaji, K. and Hso, S. (1982) Process design and simulation of absorption heat pump system. Proc. Chemcomp. – *Chemical Process and Design using Computers*, pp. 3–27.

59. Baehr, H. D. (1980) The COP of absorption and resorption heat pumps with ammonia-water as working fluid. Proc. 15th Int. Congress of Refrigeration, 1980, pp. 281–286.

60. Blakeley, R. E., Ng, D. N. and Treece, R. J. (1978) The design and development of an absorption cycle heat pump optimised for the achievement of maximum coefficient of performance. Proc. EEC Contractors Meeting on Heat Pumps, 1978, pp. 28–29.

61. Chaudharb, S. K. (1985) A study of the operating characteristics of a water-lithium bromide absorption heat pump. *Heat Recovery Systems*, No. 4, pp. 285–297.

62. Hogoard-Knudsen, H. J. (1981) Static simulation of absorption refrigeration systems, Proc. 15th International Congress of Refrigeration, 1981, pp. 214–247.

63. Kuhlenschmidt, D. and Merrick, R. H. (1983) An ammonia-water absorption heat pump cycle. *ASHRAE Trans.*, No. 2, pp. 210–220.

64. Lazzarini, R. M. and Rosetto, L. (1983) Performance prediction of a NH_3-H_2O absorption E heat pumps. Proc. 16th Congress of Refrigeration, Commission E2, 1983, pp. 317–322.

65. Schulz, S. (1972) Die Berechnung und Optimierung von Absorptionskaltemachinen Porzessen mit Hilfe von EDV-Anlagen. *Kaltetechnik-klimatisierung*, No. 4, pp. 181–188.

66. Stokar, M. and Trepp, Ch. (1987) Comparison heat pump with solution circuits. Part 1: Design and experimental results. *Rev. Int. Froid.*, No. 4, pp. 87–96.

67. Tyagi, K. P. (1987) Aqua-ammonia heat transformers. *Heat Recovery Systems*, No. 2, pp. 423–433.

68. Dykhvizen, R. C. and Roy, R. P. (1984) Numerical method for the solution of simultaneous nonlinear equations and application to two-fluid model equations of boiling flow. *Numerical Heat Transfer*, No. 7, pp. 225–234.

69. Sano, M., Kusakabe, M. and Kumi, Y. (1983) Absorption heat pump for upgrading of industrial waste heat. Proc. 16th Int. Congress of Refrigeration, 1983, pp. 55–60.

70. Buick, T. R., O'Callaghan, P. W. and Probent, S. D. (1987) Short-term thermal energy storage as a means of reducing the heat pump capacity required for domestic central heating systems. *Int. J. Energy Res.*, No. 4, pp. 583–592.

71. Engelhand, J. (1979) *Development of Resorption Heat Pump for Heating Homes.* Dusseldorf: VDI, pp. 45–49.
72. Fox, U. (1980) Anslagung einer Grosswarmepumpenlage unter Berucksicntigung verschiedene Antriesenergien. In: *Warmepumpentechnologie*, Vol. 1. Essen: Vulkan-Verlag, pp. 128–133.
73. Klein, R. (1980) Schraubenverdichter fur Einsatz in Warmepumpen. In: *Warmepumpentechnologie*, Vol. II. Essen: Vulkan-Verlag, pp. 124–138.
74. Kolbusz, P. (1974) The use of heat pumping in district heating schemes. Electricity Council Research Centre Report, ECRC/M70.
75. Mattalon, R. W. (1980) Application of high temperature heat pumps for district heating. Proc. Int. District Heat Conf., Italy, 1980.
76. Summer, J. A. (1976) *Domestic Heat Pumps.* Dorchester: Prism Press.
77. Wallman, P. H. (1987) Assessment of residential exhaust-air heat pump application in the United States. *Energy*, pp. 569–584.
78. Aleksiev, N. and Tsvetkov, T. (1983) An approach to reduce energy consumption during freeze-drying. Proc. 16th Int. Congress of Refrigeration, 1983, pp. 417–425.
79. Aublet, R. (1983) Pumpe a chaleur de grande dimension recuperation par condensation de vapeux. Proc. 16th Int. Congress of Refrigeration, 1983, pp. 67–74.
80. Berghmans, J. (1983) Heat pump fundamentals, In: *Proc. of the NATO Advanced Study Institute on Heat Pump Fundamentals*, Espinh, Spain, 1–12 Sept. London: Martinus Nijhoff Publishers.
81. Eisa, M. A. R. (1987) Heat pump assisted distillation V: A feasibility study on absorption heat pump assisted distillation systems. *Int. J. Energy Res.*, No. 6, pp. 179–191.
82. Eder, W. and Moser, F. (1979) *Die Warmepumpen in der Verfahrenstechnik.* Essen: Springer.
83. Flikke, D. L. (1957) Efficient drying using heat pumps. *The Chemical Engineer*, No. 4, pp. 592–597.
84. Gaggioli, R. A. (1978) A heat recovery system for process steam industries. *ASME J. Eng. Pow.*, No. 4, pp. 511–519.
85. Hodgett, D.L. (1976) Efficient drying using heat pump. *The Chemical Engineer*, No. 2, pp. 690–715
86. Kannoh, S. (1985) Heat recovery from waste water at drying process by absorption heat pump. *Heat Recovery Systems*, No. 2, pp. 443–450.
87. Kew, P. A. (1983) The industrial application of heat pumps. In: *Heat Pump Fundamentals* (ed. Berghmans, J.). London: Marintus-Nijhoff Publishers, pp. 260–278.
88. Laroche, M. and Solignac, M. (1975) Heat pump application to drying in agricultural and industrial fields. *Rev. Gen. Therm.*, No. 1, pp. 989–995.
89. Moser, F. and Schnitzer, H. (1985) *Heat Pumps in Industry.* Elsevier Science Publishers.
90. Newbert, G. J. and Mortin, D. J. (1983) Trends and development in nondomestic heat pump application. *Heat Recovery Systems*, No. 2, pp. 69–72.
91. Null, H. R. (1976) Heat pumps reduce distillation energy requirement. *Oil and Gas*, No. 1, pp. 96–98.
92. Peterson, W. C. and Wells, T. A. (1977) Energy saving schemes in distillation. *Chemical Eng.*, pp. 76–78.
93. Reay, D. A. (1980) Industrial applications of heat pumps chart. *Mech. Eng.*, No. 2, pp. 20–25, 310–315.
94. Reay, D. A. (1983) Heat pump research and development in the USA. *Heat Recovery Systems*, No. 3, pp. 165–170.
95. Reay, D. A. and Macmichael, D. (1979) *Heat Pump Design and Applications.* Oxford: Pergamon Press.
96. Reading, J. T. (1975) Energy consumption. Paper, stone, clay, glass, concrete and food industries. Environmental Protection Technology, Report EPA-650.
97. Van der Ree and Rostendrop, P. A. (1983) Resorption heat pumps. In: *Heat Pump Fundamentals* (ed. Berghmans, J.). London: Martinus-Nijhoff Publishers, pp. 491–528.

98. Schnitzer, H. and Moser, F. (1982) Proc. Int. Symptosium on the Industrial Application of Heat Pumps, 1982.

99. Struck, W. (1980) Antrieb von Warmepumpen mittels Dieselmotor. In: *Warmepumpentechnologie*, Vol. I. Essen: Vulkan-Verlag, pp. 26–31.

100. Supranto, S. *et al.* (1987) Heat pump assisted distillation IV: An experimental comaprison of R114 and R11 at the working fluid in an external heat pump. *Int. J. Energy Res.*, No. 2, pp. 22–23.

101. Vauth, R. (1980) Eine neue nach dem Kaltluftprinzip arbeitende Warmepumpe. In: *Warmepumpentechnologie*, Vol. II. Essen: Vulkan-Verlag, pp. 79–83.

102. Worse, S. P. (1982) Modern trends in heat pump developments. *Int. J. Refrig.*, No. 3, pp. 70–75.

103. Czekalski, D. (1985) Wplyw warunkow eksploatacji na zdolnosci chlodnicze sprezarkowej pompy ciepla. *Chlodnictwo*, No. 4, pp. 6–7.

104. Czekalski, D. (1986) Proby seryjnej produkcji sprezarkowych pomp ciepla w Czechoslowacji. *Chlodnictwo*, No. 5, pp. 8–10.

105. Gajosinski, A. and Siatka, J. (1986) Analiza jakosci procesow desorpcji w generatorach do absorpcyjnych systemow ziebniczych i grzewczych. *Chlodnictwo*, No. 3, pp. 9–13.

106. Jamer, M. and Trojanowski, k. (1984) Wplyw cyklicznej pracy pompy ciepla typu woda-woda na jej sprawnosc. *Chlodnictwo*, No. 2, pp. 9–13.

107. Maczek, K., Schnotale, J. and Wojtas, K. (1982) Odzysk ciepla za pomoca sprezarkowej ziebiarki resorpcyjnej. *Chlodnictwo*, No. 2, pp. 6–8.

108. Matheisel, M. (1985) Wplyw nierownomiernosci zuzycia cieplej wody uzytkowej w ciagu roku na efektywnosc sprezarkowej pompy ciepla. *Chlodnictwo*, No. 2, pp. 7–9.

109. Siatka, J. (1983) Waste energy recovery by absorption systems for cooling, drying and heating. Proc. 16th Int. Congress of Refrigeration, 1983, pp. 258–266.

110. Szarynger, J. and Zak, M. (1985) Efektywnosc termodynamiczna pewnej klasy obiegow porownawczych dla pomp ciepla i ziebiarek dzialajacych cvklicznie. *Chlodnictwo*, No. 2, pp. 95–99.

111. Szolc, T. (1978) Pompy cieplne. Czesc (Parts) I i II. *Chlodnictwo*, No. 3 and No. 4, pp. 1–5, pp. 18–25.

112. Urbanek, W. (1985) Pompy ciepla w praktyce. Chlodnictwo, No. 3, pp. 6–10.

113. Urbaniec, K. (1980) Sprezanie oparow w gospodarce cieplnej. *Gazeta Cukrownicza*, No. 1, pp. 134–136.

114. Wesolowski, A. (1985) Wykorzystanie ciepla odpadowego w urzadzeniu do chlodzenia mleka. *Chlodnictwo*, No. 2, pp. 8–13.

115. Duminil, M. (1976) Basic principles of thermodynamics as applied to heat pumps. In: *Heat pumps and their Contribution to Energy Conservation* (ed. Camatini, E. and Kester, T.). Noordhoff Leyden, pp. 97–155.

116. Kaushik, S. G., Chandra, S. and Gadhi, S. M. B. (1985) Thermodynamic feasibility of double effect generation absorption system using water-salt and alcohol-salt mixtures as working fluids. *Heat Recovery Systems*, No. 5, pp. 19–25.

117. Kaushik, S. G. and Kumar, R. (1987) A comparative study of an absorber heat recovery cycle for solar refrigeration using NH_3-refrigerant with liquid-solid absorbent. *Int. J. Energy Res.*, No. 4, pp. 123–132.

118. Tu, M. (1987) Thermodynamical and economic evaluation of a solar unised heat pump. *Int. J. Energy Res.*, No. 2, pp. 559–592.

119. Domanski, R. (1990) *Magazynowanie Energii Cieplnej (Thermal Energy Storage)*. PWN: Warszawa (in Polish).

120. Sokolov, J. J. and Zinger, N. M. (1965) *Strumienice*. Warszawa: WNT.

121. Paliwoda, A. (1971) *Urzadzenia Chlodnicze Strumienicowe*. Warszawa: WNT.

122. Chen, C. F. (1987) Performance of ejector heat pumps. *Int. J. Energy Res.*, No. 4, pp. 289–300.

123. *Freon* (1974) Du Pont Products Department. CH-12111, Geneva.
124. Eastop, T. D. and Croft, D. R. (1990) *Energy Efficiency* London: Longman Scientific and Technical, Chapter 6, pp. 228–230.
125. *Handbook of Fundamentals* (1972) New York: ASHRAE.
126. Ziegler, F. and Trepp, Ch. (1974) Equation of state for ammonia water mixtures. *Rev. Int. Froid.*, No. 1, pp. 30–35.
127. Schulz S. (1971) Equations of state for the system ammonia-water for use with computers. Proc. 13th Int. Congress of Refrigeration, 1971, pp. 206–215.
128. McNeely, L. A. (1978) Thermodynamic properties of aqueous solutions of lithium bromide. *ASHRAE Trans.*, No. 1, pp. 41–434.
129. Dembecki, F. and Gazinski, B. (1980) Formuly aproksymacyjne opisujace fizyczne wlasciwosci wodnego roztworu bromku litu. *Arch. Termodynamiki*, No. 3, pp. 211–224.

Further reading

Ando, A. and Takashita, V. (1979) Refrigerant absorbent pair R22-DEDGME. *Nat. Tech. Rep. Matsuhita Elect. Ind. Co.*, No. 29, pp. 102–104.

Bekheit, M. M. (1991) Selection of zeotrpic mixtures of two refrigerants as working fluids for compressor heat pumps. PhD. Thesis, Technical University of Warsaw, Poland.

Bykov, A. V., Shmilov, N. G. and Drankowski, I. K. (1982) High temperature lithium bromide absorption units for producing cold and heat. *Kholod. Tech.*, No. 7, pp. 25–31.

Eisa, M. A. R. ()1987) Heat pump assisted distillation VII: A feasibility study on heat transformer assisted distillation systems. *Int. J. Energy Res.*, No. 2, pp. 445–457.

Gruen, D. M. HYCOS: a chemical heat pump and energy conversion system based on metal hydrides. Argonne National Lab., Argonne IL, ANL-77-39.

Gruen, D. M., Mendelsohn, M. H. and Sheft, I. (1978) Metal hydrides in chemical heat pumps. *Solar Energy*, No. 1, pp. 2–15.

Jantovskij, E. J. and Janov, V. C. (1980) Isspolzovanija teploty oborotnoj vody. *Promyslennaja Energetika*, No. 3, pp. 301–310.

Kriplani, V. M., Srinivasa, M. S. and Krisha, M. M. V. (1984) Performance analysis of a vapour absorption heat transformer with different working fluid combinations. *Heat Recovery Systems*, No. 3, pp. 129–135.

Kumar, P. (1985) Experimental studies with an absorption system used for simultaneous. *Chem. Eng. Res. Pes.*, No. 1, pp. 133–141.

Latyszev, V. P. (1969) Entalpija – koncentracija dlja R22-dibuthyphyphtalate i dlja R22-dimethylether tetraethylene glikol. *Kholod. Tech.,* No. 2, pp. 61–67.

Narodoslawsky, M., Windisch, F. and Moser, F. (1980) New compression heat pump media for medium and high temperature application. *Heat Recovery Systems*, No. 1, pp. 23–31.

Pareschi, A., Gentilini, M. and Macchiavelli, P. (1983) Analysis of refrigeration fluids for heat pump system. Test plant and results. Proc. 16th Int. Congress of refrigeration, 1983, pp. 329–338.

Schwind, K. (1955) US Patent, 2938262 cl. 62-1A9.

Stal Refrigeration AB Katalog, 1986.

Stephan, K. and Seher, D. (1984) Heat transformers cycles–II, Thermodynamics analysis and optimisation of a single-stage absorption heat transformer. *Heat Recovery Systems*, No. 5, pp. 371–375.

Tomonovskaja, V. F. and Kolotova, B. E. (1970) *Freony*. Moscow: Chimija.

Wall, G. (1986) Thermoeconomic optimization of a heat pump system. *Energy*, pp. 957–968.

Index

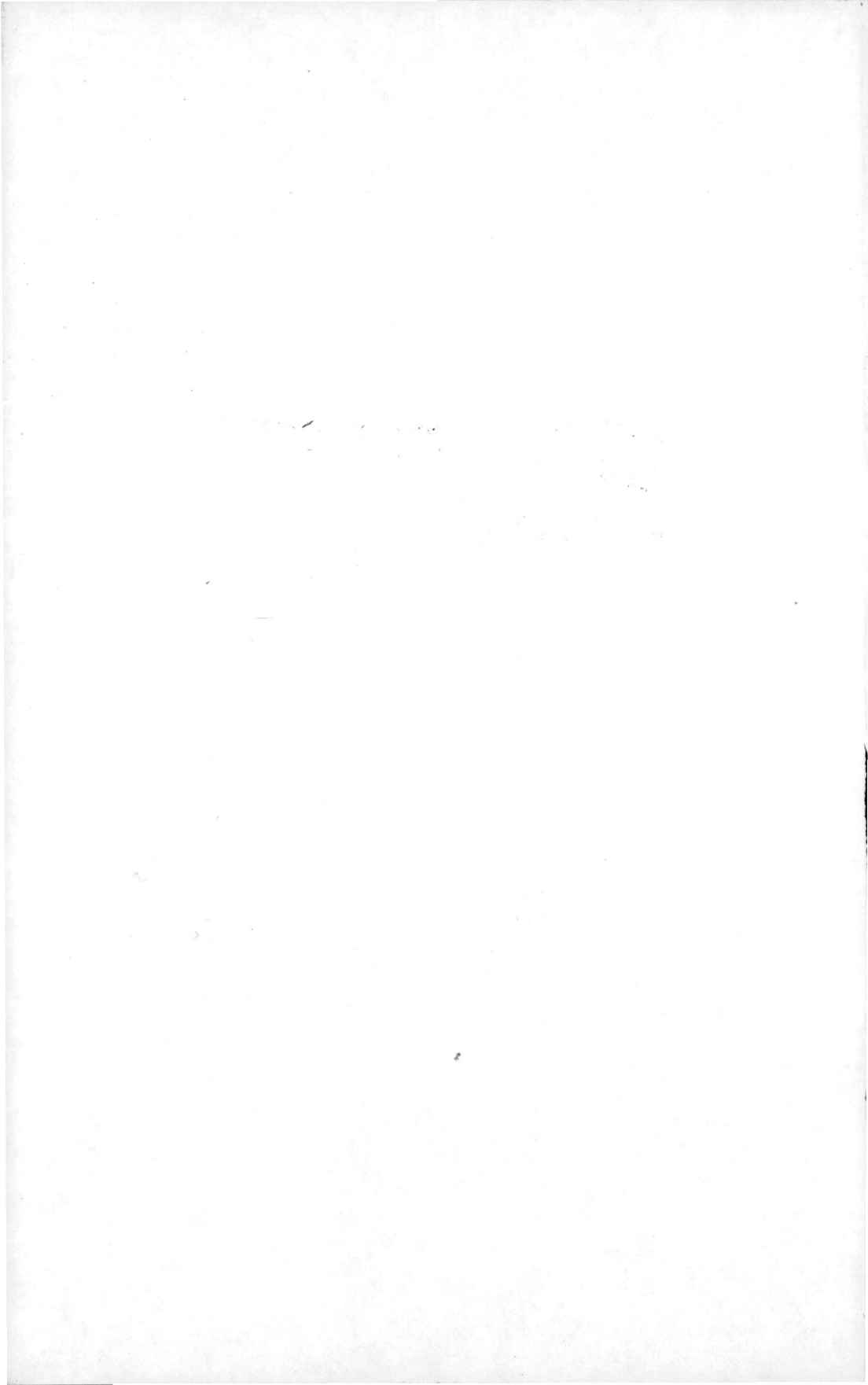